Thinking in Systems

Strategies for Problem Solving, Planning and Critical Thinking

Copyright 2018 by Alex J Golding - All rights reserved.

This book is geared towards providing precise and reliable information, with regard to the topic of cryptocurrency and its related topics. This publication is sold with the idea that the publisher is not required to render any accounting, officially or otherwise, or any other qualified services. If further advice is necessary, contacting a legal and/or financial professional is recommended.

-From a Declaration of Principles that was accepted and approved equally by a Committee of the American Bar Association and a Committee of Publishers' Associations.

In no way is it legal to reproduce, duplicate, or transmit any part of this document, either by electronic means, or in printed format. Recording this publication is strictly prohibited, and any storage of this document is not allowed unless with written permission from the publisher. All rights reserved.

The information provided herein is stated to be truthful and consistent, in that any liability, in terms of inattention or otherwise, by any usage or abuse of any policies, processes, or directions contained within, is the solitary and utter responsibility of the recipient reader. Under no circumstances will any legal responsibility or blame be held against the publisher for any reparation, damages, or monetary loss due to the information herein, either directly or indirectly.

Respective authors own all copyrights not held by the publisher.

The information herein is offered for informational purposes solely, and is universally presented as such. The information herein is also presented without contract or any type of guarantee assurance.

The trademarks presented are done so without any consent, and this publication of the trademarks is without permission or backing by the trademark owners. All trademarks and brands within this book are thus for clarifying purposes only and are the owned by the owners themselves, and not affiliated otherwise with this document.

Table of Contents

Chapter 1: What Thinking in Systems Entails — 12

Health System as Example 13

How Internal Family Systems Work 15

Therapy Within Family Systems 16

    What Multiplicity of Mind Means 17

    Exiles 18

    Protectors 18

    Example of a Man Interrupted by a Family Argument 19

The Imago Relationship Therapy 21

    How Imago Relationship Therapy Works 21

    Imago Relationship 3-step Therapy Process 22

    Benefits of Imago Relationship Therapy 23

How Appreciative Inquiry (AI) Works 24

    AI Sample Questions 25

How Different Organizations Can Share a Single System: Example
.... 26

    Example of Many Groups' Acrimony Against One Body 26

Chapter 2: The Language of Systems Thinking — 31

Terms Used in Systems Thinking 31

Attributes of an Open System 34

Chapter 3: Critical Thinking Within the Realm of Systems Thinking — 37

What Critical Thinking is About 38

How Inventiveness Goes Hand in Hand with Critical Thinking ... 40

Main System Areas Enhanced by Critical Thinking 44

    Critical Thinking Not Yet in Education 47

Chapter 4: How to Influence Matters Through Critical Thinking   49

    How to Tame Emotions Through Critical Thinking 51

    How to Know You Are Engaging in Critical Thinking 51

        Important Steps in Critical Thinking 52

    How to Use Questions to Enhance Critical Thinking 56

        What to Seek as You Design Critical Thinking Questions 58

Chapter 5: Major Strategies to Enhance Critical Thinking   62

    Foundation for Creating High-Quality Questions 63

Chapter 6: Critical Skills in Systems Thinking   69

    Thinking 70

        Constructing Mental Models 70

    How to Select Components to be Included in the Model 71

    Systems-Thinking Filtering Skills 71

        How to Represent the Components in the Model 73

    How to Develop Content-Representation Skills 74

    How to Represent Relationships Between Components 75

Chapter 7: Role of Emotional Intelligence in Critical Thinking 82

    Qualities of People with High Emotional Intelligence 83

    How Modern CEOs are Succeeding Through EI 85

        Why the Concept of EI has Gained Currency 87

        Experiment on the Impact of Emotional Intelligence 88

        Liaison Between Head and Heart for Success 91

        Why it is Possible to Learn and Acquire EI 92

        Three Main Areas Impacted by EI 92

How to Practically Enhance Emotional Intelligence 94

    Real Examples of Corporate EI Programs 94

    Best Way to Practice EI 98

    Caveat 99

    Group Practice 99

Chapter 8: Organizational Structures and Systems Thinking   101

Organizational Functions Within a System 102

Matrix Organization.. 102

    Organizational Restructuring Within Systems Thinking......... 103

    Loopholes in the Matrix Organizational System 103

    Total Quality Management (TQM).. 104

How Self-Managing Workgroups Operate 105

    How to Eliminate Bureaucracy and Organizational Politics ... 105

    Reengineering Organizations.. 106

    Changing the Reporting Structure 106

    Empowering the People Through Groups............................... 108

    Questions to Help in Group Transformation 109

Chapter 9: Introducing Systems Thinking in An Entity, the Simple Way   113

    Practical Example 115

Summarized Guidelines for Teaching Systems Thinking 117

Chapter 10: Systems-Thinking Models   118

Everything Happens to be A System 119

Individuals as Systems... 120

    Practical Examples of Bottlenecks... 122

    Problem of not Identifying and Acknowledging a Bottleneck 125

    Advantages of Identifying and Acknowledging a Bottleneck . 126

    Valuable Employees Attribute Whatever Their Specialty....... 128

    Anyone Can Identify A Bottleneck 130

    How to Identify the Best Leverage Point 130

    Examples of Working System Interventions 133

    Points to Help in Leverage 136

Chapter 11: The Balancing and Reinforcing Feedback Loops   138

    Balancing Feedback Loop 138

        Examples of Balancing Loops in Daily Life 138

    Implication of Balancing Feedback Interventions 142

    Balancing Feedback Cycles 143

        Examples Associated with Service Delivery 143

        Example Associated with Training 144

    Reinforcing Feedback Loop 145

    Installation of a Feedback Loop 145

        Reinforcing Loop Examples 147

    Best Strategies to Work with Reinforcing Loops 148

        Example of Resource Allocation in Sports 151

Chapter 12: The Process of Switching to Systems Thinking   153

    A Socio-Economic Problem Erroneously Solved Linearly 154

    Comparison Between Linear and Systems Thinking 156

    Easy Tips to Systems Thinking 157

    Learning to Distinguish Problem from Symptom 158

    Clues That What You Have is Only a Problem Symptom 159

Chapter 13: Challenges Encountered in Systems Thinking   165

    Example of Deficient System Management 165

Possible Invisible System Problems 166
Some Behavior That Mars Systems Thinking 167
Best Problem-Solving Steps in Systems Thinking 172
Main Principles of Systems Thinking.. 175
Farmers' Problems Solved via Systems Thinking 179

Chapter 14: How Systems Thinking Can Solve Social Problems 182

Characteristics of a Good Social Systems Adjustment Approach 183
Family Systems Therapy Approaches 188
Concepts of Family Systems Theory 190
Benefits of Family Systems Therapy....................................... 193
Limitations and Concerns 193
Why Systems Thinking is Best for Complex Problems............... 193
When Goals Need to Be Realigned 194
How to Deal with Feedback Delays... 194
When Some Solutions are Counterintuitive................................ 196
The Need to Put Individuals into Context.................................... 197
The Four Major System Contexts 198
Understanding People in Context 200
Principles that Guide People's Actions................................... 201
How Misunderstanding of Context Affects the System 201

Chapter 15: Appreciating People's Contexts for Profitability 206

How to Streamline Group Operations.. 207
Danger of Being Blind to the Group's Context 208
How to Strengthen Peer Groups 212
What a Robust System is 213

8

A Differentiating System 213

A Homogenizing System 213

An Individuating System 213

An Integrating System 213

How Groups Fall into Top Territoriality 214

How Groups Fall into Middle Alienation 215

How Groups Fall into Bottom Conformity 215

How to Overcome the Perception of System Blindness 217

Tips for Total System Empowerment 217

Chapter 16: How to View Your Company as a System   **219**

Important Lessons for Company Managers 222

How to Think and Interact Better 225

Five Disciplines Best for Organizational Learning 226

How to Change the Blame Game to Accountability 227

Differences Between Seeking Accountability and Blaming 228

Consequences of Blame to An Institution 228

How to Change from Blame to Accountability 229

How to Conduct an Accountability Conversation 231

How to Establish Long-Term Solutions to Problems 232

Guidelines to Help Find Long-Lasting Solutions 232

Chapter 17: How to Change Systems by Changing Mindsets   **234**

What Mindfulness Means 235

Tips to Help You Succeed in Mindfulness 237

Understanding Happenings in Context 238

Why People Resist Change 241

How Bad Habits Can be Broken via Mirror Neurons 243

Chapter 18: The Need to Consistently Think in Systems   **247**

How to Master Systems Thinking 247

How to Master the Concept of Systems Thinking 249

    Steps in the Systems-Thinking Process 249

    Crucial Systems-Thinking Skills 250

The Importance of a Management Theory 256

Why Most Approaches Fail 257

Importance of a Customized Approach 257

    Theory as an Assessment Guide 258

Managers' Roles in Success Theories 259

    Qualities Required of a Systems-Thinking Manager 259

Chapter 19: Manufacturing Company Leads via Systems Thinking   **261**

    Financial Crunch Confuses Analysts 262

    Why Big Corporations Fail to Work Optimally 263

    Toyota's Founder Acknowledges Deviation from Fundamentals 263

    Toyota Publication Explains Company's "True North" 264

    How Systems Thinking Distinguishes a Company 265

Chapter 20: Familiar Business Scenarios Requiring Systems Thinking   **267**

    Systems Thinking in Strategic Planning 267

Chapter 21: Systems-Thinking Principles Fit for The Health-Care System   **270**

    Why the Systems-Thinking Approach is the Way to Go 270

    Practical Challenges of Systems Thinking and Modeling in Public Health 272

Chapter 22: Systems Orientation – The 5 Cs of Systems Thinking 277

The Makeup of Systems Orientation 277

How the Five Cs Work in Practice 278

The Systems Thinker in a Nutshell... 283

Chapter 23: The Role of Systems Thinking in Education 284

Current Inappropriate Approach to Education 287

    Challenges That Have Hindered Transition in Education 288

Schools Operating as Open Systems 290

    Systems Impact on Education.. 292

    What the Education System Needs to Change 292

    The Look of the Recommended School System...................... 293

The Need to Revitalize Schools.. 295

    Role Played by Systems Thinking in Education Success........ 296

Sample Schools with Systems Thinking....................................... 297

Technology Experts Help Introduce Systems Thinking.............. 299

    Students Tackle Real-Life Problems 300

Objectives of Systems-Thinking Beginners Course 305

Systems-Thinking Approach in Summary................................... 307

Systems-Thinking Questions for Every Phase 308

Conclusion 309

Chapter 1: What Thinking in Systems Entails

Thinking in systems means thinking in terms of many units working together from one complex unit. This is in contrast to thinking of branches attached to a stem, where the stem can stand firm even if the branches fall off. When you think in systems, you acknowledge that the component parts of the unit are, in themselves, complete systems. As such, they should be monitored from beginning to end until they succeed, and that is the only way the big system they are part of will succeed.

In order to understand if a system is working well, whether a small system or a large one, it is important that you monitor it over time. That way, you are in a position to tell how one action is affecting other actions, and how, in turn, they collectively affect the way the entire system works.

Thinking in systems works in contrast to the traditional method of analysis, where one takes a system and breaks it down into different segments for ease of understanding. In that traditional method, one can contemplate doing away with one segment if it proves too costly, but in systems thinking one considers how to make each of the constituent systems cost-effective, as it is a crucial part of the bigger system.

Thinking in systems is not an academic issue but one that is applicable everywhere in everyday life. For example, the medical sector is a system, as is politics. Even the education sector and the economic sector are systems. These, and many more, are systems that operate in a country, which is itself a system. In fact, different countries are systems that comprise the complex global system.

Some experts like to look at the universe as one massive system comprising almost an infinite number of systems, all of which operate under similar principles.

The main idea in writing this book is to show people how systems operate, and what every individual and organization can do to make a system work efficiently for the good of the bigger system it is part of. And since, as has been explained, our entire life is about systems, systems thinking is, therefore, about how to make life more manageable and, ultimately, better.

To expand on how a system looks and operates, let us take the health system as an example.

Health System as Example

In order to apply the principles of systems thinking in the health system, you need to ask yourself: what is required in order for the system to keep running well? You will then find that, first of all, health is necessary, and it is also important that, once health has been attained, it is sustained. In order to have a health system that runs well, you will need to look at its health and sustainability as core goals. The next question could be: what do I need to address in order to attain health, and what do I need to do to ensure good health is sustained?

Before you can establish what is important to address in order to attain health, you need to understand health itself within the context of a human system. Health can be explained as the ability of the human system to meet the needs of a human being. From that point onwards, you can then begin to analyze what those needs are and what systems are in charge of addressing them.

Some of the systems you will realize are basic to human needs include food supply, health and social systems, among others. Once people are free of disease, they eat well, are socially comfortable and are generally happy. So what are the systems you need to maintain in order to ensure that status continues? You need to ensure people's financial systems work smoothly, their social networks impact their lives positively and so on.

Since sustenance has time as a basic element, you need to look at how the systems work to prevent anything from disrupting health along the way. In short, you will be looking to preempt any danger, including physical and psychological trauma, while strengthening defense mechanisms. In order to strengthen defense mechanisms you will need to address them within the smaller systems associated with the human system. In this regard, being vaccinated preempts disease, while having saving accounts and various forms of insurance serve as the financial system's defense mechanisms.

Addressing system issues, and by implication, dealing with systems thinking, is something everyone does on a daily basis, whether we realize it or not. It just so happens that the actions we take naturally serve their purpose within the context of their respective systems. For example, while eating well and exercising enhances your health system, building a career enhances your financial system. Yet those actions that we take as a matter of course improve various important systems, and since those systems happen to be interlinked, the entire complex system is improved.

Take, for instance, the benefit of having a successful career. It puts you in a position where you can afford good food to improve your immune system and can pay gym fees that enable you to exercise and strengthen your entire physical body system. This is an example of how systems work in tandem with one another for the success of the bigger system; in this case, the human system.

In trying to understand and improve a person's welfare, the main fields experts usually analyze are:

1) Internal family systems (IFS)

2) Therapy provided within family systems

3) The Imago Relationship Therapy

4) Appreciative inquiry

How Internal Family Systems Work

To understand internal family systems well, you need to first appreciate that there is an original whole self, a complex system whose constituent smaller systems work efficiently and in harmony so that you are a happy, confident person, able to show compassion to others. However, things happen in life that mar some systems that are part of yourself, and the entire system can change for the worse.

When you undergo a painful experience, such as losing a job under unfair circumstances, you get angry, and sometimes you have no words to explain to people at home how that happened, so you feel ashamed. Consequently, you stop being the same happy person that you used to be. You may even begin, for whatever reason, to believe envious neighbors are pleased you lost your job, and that perception may drastically erode your self-confidence.

Thus, from a confident, happy person, you will change to a self-conscious, withdrawn person, and that is because your emotional and mental systems have been adversely affected. In the field of psychotherapy, experts consider a person's whole system as comprising many sub-personalities, which, in our context, we can consider to be smaller systems. How do you then recover your wholeness and become the happy, confident person you used to be?

How does a person who is prone to anger, or one who is too sensitive to criticism, learn to become relaxed and open-minded? How does one shed extreme jealousy to become a person who can embrace other people's success with joy? These negative feelings are injurious to a person's mental system, just as malnutrition is harmful to the physical system, and they need to be addressed in the same way a person takes up healthy eating in order to become physically strong again. In fact, the analysis of the state of health of a person's subsystems is the purview of the IFS, and then what follows is the subsystems' restoration to good health so that, at the end of the day, all the subsystems are in balance and working in harmony with one another.

It is important to cover the entire scope because that is what constitutes a person's overall health.

Another way to understand IFS is to look at a person as having an ecology of minds, all isolated from one another, and every one of them possessing valuable qualities. In short, every one of those minds, no matter what they urge you to do, wants to take care of you. Some may be confused by this latest assertion, considering that not everything some minds advise is noble. Nevertheless, this can be understood in the context of the life you lead.

Therapy Within Family Systems

Sometimes life experiences force some parts of you to abandon their noble roles and to restructure your system in an unhealthy manner. It is easy to appreciate this scenario if you think of a family afflicted by alcoholism, where children in that family adopt protective roles courtesy of the family's extreme dynamics. However, using the example of a child who is always the family scapegoat, or one who looks perennially lost, you cannot conclude that what they do or how they appear is who they essentially are.

Instead, you need to appreciate that if you remove the children from the alcoholic environment and undertake further intervention so that the kids resume their natural roles as children, the unique talents of each child will show up, making the child an individual separate from the chaotic environment that created him or her.

This analogy can be applied to a single individual, where external circumstances force certain parts of the self to take up extreme roles. Like the lost child, once things are normalized, the parts that had originally taken up extreme roles resume their valuable roles, normally laden with compassion, and the self is back in balance.

IFS has been the subject of study for some time now, and experts like Dr. Richard Schwartz have devised ways of working with individuals, couples within marriage and even whole families. What these experts have done is consider the benefits of the traditional model used by other authorities, along with the multiplicity of the mind, and incorporated them into systems thinking. They have, therefore, ended up with a comprehensive approach to solving problems.

What Multiplicity of Mind Means

It should be easy to appreciate multiplicity of the mind if you think of times you have had conflicting ideas about something. Even an issue that can ordinarily seem black and white sometimes becomes contentious, not because you are debating someone else about it, but because different opinions crop up in your own mind. If you consider your day-to-day life, you may be able to identify instances when you wanted to watch a movie or a TV show, but, at the same time, you remembered an incomplete project which is already due. Those two lines of thinking emanate from two of the minds that you have.

How many times have you told yourself, I have some urgent work at home but I don't want to leave this party? Or, according to my doctor I shouldn't be drinking, but I don't want my friends thinking I've changed? All these kinds of divergent voices represent things that you hold dear: morals, goals and values. Your moods and attitudes are also brought into play by the different minds that comprise part of who you are.

The interesting thing is that, as long as you are in good health and things are going well for you, meaning you are happy about yourself and whatever is going on in your life, you hardly have these conflicts of minds. It is like there is no multiplicity of minds within you. This is another way of saying that, as long as you are in a happy place, all the different systems within you function smoothly in a way that reflects on the well-being of your entire system.

If you contrast that with the times when you are feeling anxious, you will notice you have trouble making up your mind on what action to take or how to respond to various situations. That is how you find people doing things that are very much unlike them, only because one of the voices was more dominant than the rest that are, possibly, more typical of the person. The emotional intensity a person experiences can end up pushing one or more parts of the person to the extreme, and that is the person you end up seeing. Sometimes people do things they end up apologizing for, and that is because, although they knew better, they ended up acting differently, probably due to those conflicts of minds.

People's behavioral trends are predictable with regard to some parts moving to the extreme. There are normally two alternatives: those parts either step in and influence matters, or they back away and hide. It is also possible for parts that do not step in to be thrust into exile as the internal system seeks to stabilize matters, and when that happens, those parts are referred to as exiles.

Exiles

Exiles, while hidden, have intense feelings, and they can be triggered by events of a highly emotional nature. They are normally linked to events that happened in the past, which a person has not been able to confront and work through.

Protectors

Those parts of a person that take charge, including helping to make decisions, are referred to as protector parts, and they come in two categories.

(a) Firefighters

Instead of hiding or going into exile, these protectors react to triggers and actions, especially when things seem to be getting out of hand.

(b) Managers

Sometimes the protector parts act proactively, anticipating events and acting in readiness. Those are the situations when you may find a person behaving in a calm manner even when something of great magnitude has happened. These protectors can manage things so well that observers might not even realize there's a near-crisis.

Does this mean that people's inner personalities manifest summarily as exiles, managers and firefighters? No. There is more to the inner person than those three categories, but those other parts usually interrelate well and work in harmony apart from the occasional times when some are pushed to the extreme, finding themselves in conflict with others.

Example of a Man Interrupted by a Family Argument

Take the example of a man, John, who is reading and hears his wife and son arguing in the living room. His wife is admonishing their son for not tidying up his room and the son is trying to explain himself. John has two managers; one that wants him to rise and go shut the door to the study so that he can no longer hear the argument, and another that wants him to go right into the living room and try and sort out that argument.

John has a boy part that empathizes with his son, recalling how humiliating it is to be admonished by a parent. John also has an adult part within him that empathizes with Jane, imagining her frustration at being unable to get their son to do his chore. The two parts at work here are exiles, and they would rather Joe left the situation alone. Then there is the firefighter in him, urging him to go tell off both his wife and his son for interrupting his peaceful study.

The fact is John loves these two people who are annoying him. How do you think he is going to react to the situation? Really, how John reacts will depend on which part of himself he pays most attention to. Another question then arises: how will he choose the part to pay most attention to?

There is a conflict among several parts of John that he must resolve. It is important to realize that John cannot solve that personal conflict by shoving the urges of one or more parts aside. He has to acknowledge and appreciate the existence of those feelings because it is a fact that they are being triggered and prodded to extremes.

Specifically, John has to:

- Acknowledge the existence of the manager that wants him to close the door to avoid the noise, and also appreciate that this manager is pushing him in that direction for his own welfare.

- Acknowledge the existence of the manager that wants him to go settle the ongoing argument between his wife and his son, and also appreciate that the reason this manager is pushing him towards that direction is to bring harmony back into the home.

- Acknowledge the exile that is reminding him of childhood scoldings and accompanying humiliation, even as he acknowledges identifying with Jane's position of feeling helpless. He must then appreciate that both scenarios cause him a certain amount of shame.

- Acknowledge the firefighter part in him and appreciate that it is pushing him to take advantage of the opportunity to vent the entire day's anger and frustrations, and not just the anger of the moment brought about by the conflict between his wife and son. John also needs to appreciate what it will mean to adhere to his firefighter self and pay attention to the part that allows him to choose a better way to meet the prevailing needs.

To appreciate the different parts of himself, John has to observe them from a position of identity that is entirely different from those parts, and also use a lens that enables him to see all the parts with clarity. He needs to engage his internal awareness, above the separate parts of himself, which is capable of looking at his various minds with compassion. Another way of looking at John's position as he seeks to handle the situation is that his internal self is the foundation upon which other divergent parts of him emerge. Hence, the self has a way of handling the influence of each part while ensuring each part remains intact.

The Imago Relationship Therapy

While still dealing with the aspect of human health, it helps to look at how relationships between one person and the next can be enhanced, including relationships between couples. This is particularly important because all systems within the systems-thinking model work with and around people. The overall health of individuals is, therefore, central to the success of virtually all systems.

Imago Relationship Therapy, a relationship model originated by Harville Hendrix, a psychologist, and his wife, Helen Hunt, is applicable to systems thinking as it deals with relationships between individuals, showing reasons why certain individuals are attracted to each other, why the relationships they form are important and what to do in order to sustain those relationships.

According to Hendrix, people have an internalized image of the kind of person they want to engage with in a romantic relationship, and that emanates from their childhood experience. In an unconscious way, they want to relate with people who resemble their childhood caretakers, and unfortunately, they do not only seek out the caretakers' good traits. Instead, they want people who are like those minders they had in childhood, warts and all.

Due to this indiscriminate attraction to certain people, problems usually arise that trigger negative issues from childhood. What the Imago relationship model acknowledges is that the attraction to a person's negative traits, as experienced in childhood, is a reflection of a silent desire that a person has to be healed. There is a part of a person that is apparently still stuck in childhood, and so he or she longs for healing in order to become whole and mature. Imago therapy, which is particularly common in couples' therapy, is designed to help individuals heal so that they can attain their life goals.

<u>*How Imago Relationship Therapy Works*</u>

Imago relationship therapy seeks to:

- Eradicate negative language that is usually hurtful.

- Create an environment that is safe and conducive to mutual communication.

- Eradicate strong imbalances that gain prominence at the expense of an individual's feelings.

Sometimes the technique of behavior change used involves a dialogue where partners can channel the negative emotions within them. The partners get a chance to communicate these negative emotions that are associated with experiences from childhood.

For this technique to be effective, the partners frame their general needs in the form of small requests that are easy for either partner to fulfill. When couples are being counseled, as one partner prepares to put forward a request, the therapist advises the listening partner to view the anticipated request as a gift. The idea here is to have a positive dialogue between partners, which is also loving and mutually respectful. Psychologist Hendrix prefers a three-step process to nurse a suffering relationship back to health.

Imago Relationship 3-step Therapy Process

The three steps provided by Hendrix for the process of Imago relationship therapy are:

(i) Mirroring

(ii) Validating

(iii) Empathizing

Remember that the reason you want to undertake Imago therapy is so you can have a great relationship with the people you are associating with, whether at home, in the office, in your business or anywhere else. For that reason, it is crucial that you keep an open mind as you go through the Imago exercises.

When you are mirroring, it means you are listening to what the other person is saying and repeating it, and as you do that you shun any preconceived ideas about the other person's viewpoint. In short, you need to listen with an open mind and repeat what the other person has said without judgment and without being clouded by your personal interpretations. When you repeat what the person has said in that manner, you are actually acknowledging that you are taking the person's opinions into consideration.

In the next step, which is validation, and which comes after you have acknowledged the other person's views, you need to validate the other person, as well as the person's contribution, even when you do not agree with the message.

Once you understand what the other person has to say, and acknowledge it and validate it, you need to show empathy. This means that you try to understand the other person's emotions and feelings, and it is important that you formulate your response carefully, for clarity. When the other person fully understands that you appreciate their viewpoint, it reduces the chances of misunderstanding and miscommunication.

Benefits of Imago Relationship Therapy

Imago relationship therapy helps improve communication between partners, and the way they relate henceforth becomes a great source of fulfillment. Many couples are able to shed their negative feelings, including those resulting from unfortunate childhood experiences. Even those who have harbored self-hatred are able to overcome it. At the end of the day, the couples are able to sustain a healthy relationship based on mutual love.

People can even undergo Imago relationship therapy as a group, solely to strengthen their relationships. According to Hendrix, the main goal of undergoing Imago relationship therapy is to foster unconditional love, not only for the individual but also for other people the person relates closely with. After undergoing this therapy, individuals learn how to resolve grievances without judging or antagonizing one another.

Imago relationship therapy should not be mistaken as therapy for spouses only. In actual fact everyone can benefit from this form of therapy. Following the same principles used by couples, one is able to do a self-examination, and in the process learn how to respect other people. One also learns how to give unconditional love and commit to a partnership.

How Appreciative Inquiry (AI) Works

Appreciative inquiry (AI) is a technique that uses change management to focus on identifying areas where the system is working well. It then analyzes those areas and, whatever is going well, management commits to doing more of it. The basic principle on which AI works is that those areas which people in an organization emphasize are the ones that grow. Thus, by emphasizing your own chosen areas, you are setting a direction for your organization. Through AI, you have the ability to drive an organization in the direction you please.

In order to understand AI better, think of the converse—focusing on the problems. If you put your time and resources into identifying problems, then you concentrate on finding ways to solve them, the organization will excel in problem identification and solving. On the other hand, if you invest your time and resources in identifying the organization's strengths, and then you concentrate on strengthening those strong areas even more, the organization will excel in identifying the organization's strengths and building on them. The latter represents what appreciative inquiry is about.

How do you find out what an organization's strengths are so that you can focus on them? You need to use guided questions of a certain kind, so that you can encourage everyone involved to think positively, and you can also enhance interaction among employees. Those guided questions cover four main areas that help you discover, imagine, design and finally, deliver (DIDD).

AI Sample Questions

(a) Discover-type questions

These questions make it possible to identify processes that work well within the organization.

Question: What aspect of the service launch do you think went very well?

Answer: One thing that came across as very successful was using Facebook because word about the launch spread like wildfire and triggered great anticipation.

(b) Imagine-type questions

These questions make it possible to analyze why certain processes are working well, and they set the stage for brainstorming in order to gain knowledge that can then be used in other processes.

Question: Why was Facebook successful as the service launch tool, and how else could it probably be used to bring such success?

Answer: It was great because of how easy it was to carry out. It required very little time and did not cost money. It was also exciting and enlightening to see people discuss the service long before the commercial launch. It is possible Facebook could also be used to promote the seasonal service price offers.

(c) Design-type questions

These are questions that enable the people involved to design a course of action—an action plan.

Question: How can we go about testing promotion of seasonal service price offers using Facebook?

Answer: The company can make a special offer for next month, because it will be low season. Liam can advertise that offer on the company Facebook page while other employees with personal Facebook accounts post the news of the offer and tag their friends.

(d) Deliver-type questions

These are the questions that enable you create a criterion by which you can measure the success of your actions. They help create a way to determine if the executed action plan has led to success.

Question: How shall we determine if it's worth Liam or other personnel's time to post the special offer on Facebook?

Answer: We'll have at least one more customer each day, over and above the minimum daily customers for the season.

It is important to keep in mind that, just as systems thinking is applicable in all spheres of life, so are its subsidiaries such as this AI. You also need to note that the goal of AI is to help organizations improve on whatever it is they do best, building on their strengths in a positive manner. For that reason, people need to customize the four major areas of discover, imagine, design and deliver in such a way as to meet the needs of the individual, family or institution.

How Different Organizations Can Share a Single System: Example

Example of Many Groups' Acrimony Against One Body

In 1999, when the third World Trade Organization (WTO) ministerial meeting was held in Seattle, USA, there were massive demonstrations. The interesting thing was that, although there was one single opponent, the complaints were as diversified as the complainants. Members of certain non-governmental organizations (NGOs) demonstrated because the WTO's policies allegedly suppressed developing nations while others demonstrated because the policies favored big corporations, some of which were responsible for environmental degradation. There were even demonstrations by America's Green Party environmentalists and people who felt the

WTO was responsible for the unfair distribution of health care in Western countries. Some people demonstrated as crusaders of animal rights, others in opposition to unchecked globalization and there were a host of other complaints, many against mega-corporations. How can you relate this to systems and systems thinking?

Well, the diverse groups are systems that are complete on their own, and they have their unique interests. The WTO is a system, too, separate from the others and much bigger than the rest. Yet the way the WTO operates affects the other systems both positively and negatively. Some of the systems are affected more negatively than others, but the fact remains there is some kind of symbiotic relationship between each of the systems and the WTO. As an example, the system forming the ultra-wealthy mega-corporations was apparently affected very positively by the WTO's policies. On the other hand, many of the other systems represented by NGOs, animal rights activists, environmentalists, individuals and others felt negatively affected by the WTO's policies.

In order to conclude that systems are interrelated, it is important to consider what they do and how their actions affect other systems. In the case of the WTO, the organization influences the parameters of transnational commerce, and so, if it is unfair in how it goes about its work, some countries are bound to be negatively affected. Consequently, trade organizations, individual traders and manufacturers within those countries are negatively affected, and everyone else involved with them feels the brunt of the WTO's actions.

One thing that becomes clear, even to those not concerned with what the WTO does, is that the organization, though trade-based, affects people's everyday lives. It is therefore safe to say that even a trade system can affect a social system. Therefore, systems whose actions affect one another need to work together in order for the health of each to be sustained. Though individual companies may be business-based, they are systems whose actions and manners of operation affects social institutions, including individuals, families, communities, countries and the world at large. In fact, as long as companies engage people, sell to people and continue to associate with people, it is correct to consider them social institutions, and their modes of operation inevitably affect people, both individually and collectively.

To expand on the relationship between individuals and companies, consider how a company employee tries to balance family and work life. Think about how people anticipate benefitting from ongoing research in biotechnology. Do people in developing countries not shudder when big companies dump their manufacturing waste along their shores? When telecommunication companies lower their tariffs, is it not a big benefit to individuals who enjoy low communication charges? In short, whether a company operates locally or globally, it is a system that affects the systems of individuals and human networks on a daily basis.

Conversely, individuals affect a company's health and level of success. So, the relationship is not one-way. If a nation's population is highly unhealthy, either physically or mentally, employees cannot contribute optimally to the success of the company. Also, if they are generally not well educated, the same situation prevails. In short, whether systems are big or small, when they interrelate, the positive aspects of one affect the other, and the same case applies with the negative ones.

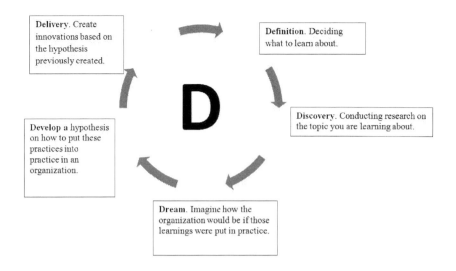

When one considers how one's actions affect another, one important thing that cannot be ignored is how some systems are more powerful than others. For example, if the population of its resident country is poorly educated, a company can decide to outsource work. But if a company sells its unique products for exorbitant prices, there is little or nothing local consumers can do, especially if the cost of importing alternative products will be higher.

For this reason, the stronger system needs to take responsibility to ensure it does not act selfishly, ignoring the interest of other systems. As shown by the 1999 WTO demonstrations in Seattle, if a big system does not take responsibility and is insensitive to the needs of smaller systems, the smaller systems begin to resist the pressure. In the case of the WTO demonstrations, these were a way to communicate the demand for the big system to become sensitive to the needs of other smaller systems. The demonstrators wanted the organization to begin taking responsibility for its actions.

For a big system to behave positively towards other smaller systems, it needs to have what some experts have called "response-ability." This means the ability to acknowledge the things that affect people and design ways to address those concerns. The system also needs to be open to learning and making changes. One might wonder why there was not much talk some years ago about systems thinking, particularly about the necessity for corporations to think about people outside their own organizations.

The reality is that things have changed, and with continued globalization, even people from countries that were previously marginalized now understand their rights and continue to be emboldened to demand them. A company that dumps waste carelessly in an area risks the wrath of the surrounding community even if the government turns a blind eye. As such, even as companies seek opportunities, they also need to check for danger signs and take action to avert calamities. Sometimes the required actions may be radical, and a company destined for success must have the ability to make sweeping changes to its processes when these become necessary.

Chapter 2: The Language of Systems Thinking

Systems thinking is not a technical subject as it involves activities and everyday happenings. Nevertheless, systems thinking has not been widely used as a management or problem-solving approach for long, and so its details might sound new to many people. For those who have heard of systems thinking before, some terms may be very familiar. However, some may still not be clear on certain terminology.

In this chapter of the book, you are going to learn various terms commonly used in systems thinking, as well as their meaning in context. Once you know these terms, you will be in a position to clearly understand the information given either in this book or elsewhere, thus avoiding the risk of misinterpretation.

Terms Used in Systems Thinking

(1) System

System refers to a set of different elements functioning as one system with a view to achieving a common goal.

(2) Subsystem

Subsystem refers to one component of a system that is much larger than the one being defined. Good examples include the circulatory system, a subsystem of the human system, or the railway system, a subsystem of the transport system. At times, the system that subsystems are a part of is termed supra-system, but that is particularly when someone is speaking about it in relation to the subsystems that comprise it.

(3) Element

An element is a system's component and is necessary, but by no means self-sufficient. In other words, for the system to accomplish its purpose, it must have the element, but the element on its own does not have the capacity to replicate the functions of the system.

(4) Synergy

Synergy refers to the interaction among the different systems that work in tandem with one another to form one overall system. The reasons systems work harmoniously is because of the synergy in the entire system that comprises smaller parts, and the relationship among those component parts is the one that adds value to the system.

What actually maintains the relationship that exists among subsystems, or the component parts of the system, is energy they exchange from one to another and vice-versa. That energy is the binding factor among elements in a system, just as money is the binding factor among people in the banking sector. It binds as effectively as heat binds elements in a thermodynamic system or as a learning system is bound by information. Thus, without that binding factor there is no whole. Without synergy there is no system—just an assembly of elements.

It is important to know that what maintains the relationship among system elements is the difference of energy potential among the elements involved, and this enables interchange. As a matter of fact, as long as a system is healthy, it will continuously search for a dynamic balance, and this search will go on through self-regulating systems. A human system, for example, has the capacity to maintain body temperature in a dynamic balance of around 98.6° Fahrenheit. It increases or decreases blood circulation in the areas close to the skin, the skin serving as the system boundary, and that is how you experience shivering, perspiration and even panting.

The total amount of energy a system has is fixed, although it is constantly redistributed among existing subsystems.

Entropy

Entropy is the process through which energy is distributed all through the system in an even manner. It is important to know that every system and subsystem experiences entropy. Note that if the energy level among the different elements or subsystems is zero, the system will inevitably collapse, fall apart or die, depending on the nature of the system.

Essentially, therefore, a system imports energy across the boundaries of constituent subsystems, or across the boundaries of its constituent parts, in order for it to continue its existence. Alternatively, it can continue to exist if it can create new energy sources.

Open System

When a system has the capacity to import energy, or even to export it across boundaries of its constituent parts or subsystems, it is referred to as an open system.

Closed System

A closed system does not have the capacity to import energy. If ever a system is unable to internally generate enough energy to replace the energy lost though entropy, it is, inevitably, destined to die.

There is a great real-life example in the demise of the Union of Soviet Socialist Republics, popularly known as the USSR. The USSR was a closed political system that was unable to maintain itself. If only it had had the capacity to generate enough energy internally, or it had had a way of importing necessary energy, it would have survived. Instead, the states that comprised the USSR—Georgia, Yugoslavia, Czechoslovakia and others—began to break off one by one, and now there is no longer USSR, a system that was once a massive global political force.

To use another example, you can now probably understand why it is extremely difficult to solve educational problems by way of general solutions. You need to consider that the educational system comprises numerous sets of elements, all of them unique in their own ways, and they are particularly organized in a constellation of relationships, each of them unique.

Attributes of an Open System

After learning the definition of open system, it is now time to learn how an open system operates. Schools, which are essentially social systems, are considered to be open systems. In their 1966 publication, Katz and Kahn described the attributes an open system has. Those attributes are listed here, together with some explanations on how the system basically works.

• Energy within the open system is transformed, producing an entirely new thing.

• The environment receives a product exported from the open system.

• The system has a cyclical energy pattern, and the product the system exports into the environment is the energy source responsible for the repetitive nature of the activity cycle.

• The system works towards maximizing its ratio of imported energy to expended energy.

• An open system manifests differentiation, which is a tendency to intensify complexity via specialization.

• Open systems, obviously, have a high level of openness, and in addition, they are characterized by the concepts of hierarchy and purposiveness as well as homeostasis.

Hierarchy

A hierarchy is a level system where the base level is the lowest, and every successive level is higher than the previous one. Wherever there is a successive level, the one below it has fewer processes and is therefore less complex. Every successive level has the processes and complexities of the lower level, plus others of its own. Consequently, the relationships involved become more complex with each successive level, because elements continue to increase in number accordingly.

As the elements or subsystems continue to increase in number in a linear fashion, the relationships, too, continue to increase in number exponentially. As this incremental growth occurs, more energy is needed at a faster rate to maintain the increasing relationships.

Hierarchies come in different forms, and some are even natural, such as the order in which children in a family are born. Some hierarchies are arbitrary, for example, a school system that has been specifically designed. Natural hierarchies do not need as much energy to maintain as arbitrary ones. In fact, arbitrary hierarchies often divert energy from the intended goal.

One lesson that needs to be learned from expert Kenneth Boulding is that, in hierarchies, people often tend to look for the cause of problems in the wrong place. For instance, where a teacher seems to be encountering management problems in a class, people may criticize the behavior of the teacher, meaning the teacher is being blamed at an individual level. However, the problem at the classroom level might not lie with the teacher at individual level.

Very likely, the solution can be found at the school level, coming in the form of supportive structures within the school system. One suggestion provided by Russell Ackoff is that the best way to handle a problematic situation is to dissolve it. There is a high probability that the individual teacher is encountering classroom management problems because of an arbitrarily organized instruction where self-contained classrooms are teacher-managed.

Homeostasis

Homeostasis is a very important characteristic of a system and has the connotation of self-regulation by way of feedback mechanisms. For example, machines happen to be relatively simple systems, and they also have few variables. Moreover, these systems function in a relationship that is generally stable and which require very little feedback from their environment in order to function.

The converse is true of organic systems. They are extremely complex, for the most part, and they have a wide range of variables. In addition, those variables require massive feedback, which is energy expended. The bigger and more complex an organic system is, the more energy it requires in order to have the capacity to maintain a dynamic balance among its many elements.

Purposiveness

Purposiveness is the remaining system characteristic. As Bánáthy points out, there are some systems that have a very clear singular goal. Such systems are categorized as being unitary. Those systems that have numerous goals, sometimes conflicting, are referred to as being pluralistic.

Chapter 3: Critical Thinking Within the Realm of Systems Thinking

Traditionally, there have been two approaches towards excellence, one being the holistic approach, where the world is seen as a universe with elements within it that are interdependent. The other is pluralism, where the world is seen as being a multi-verse, with varied perspectives towards reality. In the latter case, scientist Humberto Maturana says controversies over the many contradicting views would cease if everyone always operated with "objectivity in parenthesis." However, other experts have found an even better way, one that marries the holistic and pluralistic approaches. It not only introduces grounding and purposefulness to everyday operations, but also injects a sense of responsibility within systems thinking.

According to Professor Jake Chapman, who has written extensively on systems thinking, the crux of systems thinking is having a good visual of the bigger picture, one that transcends abstraction and appreciates perspectives from other people. In reality, therefore, whether it is a case of an individual, family or even organization, one needs to be able to step back when a situation does not look good, disengage from the mess, complexity and accompanying uncertainties, and then seek clarity on the main interrelationships involved. In the process, one obtains different perspectives regarding the situation.

For the process to be fruitful, it is imperative that one acknowledge several perspectives from stakeholders, even when those perspectives are contrasting. That is the best way to engage in informed thinking and to connect that thinking with the best course of action in order to create improvements that are not only profitable but also morally justifiable. This process falls under the realm of critical thinking, so it is important to understand how to go about it.

What Critical Thinking is About

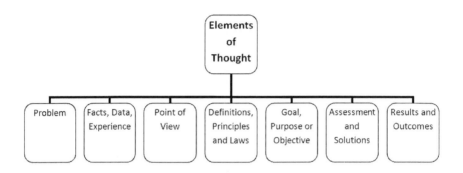

The process of critical thinking entails analyzing issues and evaluating them with objectivity, because at the end of the day you want to make a credible judgment of the situation. To ensure effective critical thinking, there is a simple procedure you can follow:

•   What concepts are involved in this situation? How do the individual concepts relate to one another?

•   How relevant are these concepts to the situation at hand? In short, what is the importance of every one of the concepts in the situation you are trying to analyze?

•   What are the arguments presented in this case? Identify and evaluate them. Note that identifying and analyzing present arguments is not sufficient. You need to be able to design your own arguments that show what you think of the other arguments presented. Do you agree or disagree with those arguments?

•   Arguments, by their very nature, take you through some form of reasoning. It is important that you be on the lookout for any inconsistencies therein, as well as any glaring mistakes in the premise and overall reasoning.

- Identify problems in the situation and try to solve them in a systematic manner.

- Consider your own beliefs and values and see if they are justified as you proceed with the process of critical thinking.

From these guidelines, you must have noted that there is more to critical thinking than amassing knowledge. Getting information is important, but it must not just be random facts. You need to ensure the information you gather is relevant to the situation under consideration, so that when you present your arguments, support existing ones or question them, you are able to knowledgeably support your position. Casual arguments have no place in critical thinking. All assertions and criticism need to be credibly supported. There are many people who have access to information, but only a fraction of them make an effort to engage in critical thinking.

Yet, without critical thinking, it is not possible for anyone to effectively think in systems. When you engage in critical thinking, you become logical, are able to link the facts you have gathered with possible consequences and can apply the knowledge you have acquired to solve the identified problems. This is another way of saying that when you know how to go about critical thinking, facts and information become tools to achieve bigger things.

As a person adept at critical thinking, you need not shy away from making arguments to support your position. As long as you are arguing from a position of knowledge, people are likely to pay attention to you, and you will have a good chance to influence them for the better. On the contrary, people who make arguments without relevant information often come across as argumentative, with a penchant for criticizing other people and other people's positions. As has already been made clear, every single system in the universe has a way of affecting other systems, and so, in order to make improvements to one system, one has to consider the role of other systems close to it.

In the course of critical thinking, be prepared to support workable arguments and to oppose those that are not factual. What a critical thinker seeks to achieve at the end of the day is:

• To detect and expose any inadequate reasoning in arguments presented.

• To identify and expose any fallacies in the case.

• To be a great contributor in mutual reasoning.

• To contribute positively to constructive tasks.

• To employ the information and knowledge gathered to make improvements on the theories in use.

• To employ the information and knowledge gained to make good arguments made even stronger.

• To use the ideas and knowledge gathered to enhance the working processes already in place.

• To utilize the ideas and knowledge gathered to enhance social institutions and make them stronger and more effective.

How Inventiveness Goes Hand in Hand with Critical Thinking

Those encountering critical thinking for the first time, seeing that emphasis is placed on well-supported arguments, may fear that their creativity might be inhibited by the process of critical thinking. However, there is good news in that creativity fits in very well within critical thinking. Even when you are being creative and want to introduce new approaches to solving a problem, you cannot help but critically evaluate the situation and the approaches you potentially want to employ. Only after critical evaluation can you reasonably decide that you have a good basis upon which to improve on existing ideas. In short, you do not have to shove logic aside to be creative.

Granted, the critical thinking process helps to keep people rational and logical, but it does not bar creativeness. On the contrary, it is during the process of critical thinking that you are able to see the possibility of solving the problem at hand through unconventional means. The process enables you to think broadly, outside the box. You will be able to confidently challenge consensus and take positions that may not necessarily be popular.

Think about a new company being established in a rural area where incoming and outgoing suppliers, customers and other stakeholders are bound to disrupt the otherwise quiet life of people. Will it be sufficient for the investors to acquire legal documents allowing them to establish the company in the locality? Definitely not! It will be advisable to think beyond the legalities of being in the locality and to consider the potential relationship the company management, employees and everyone else involved will have with local residents.

For a business enterprise, whatever its size, to thrive and excel, it requires a peaceful and friendly environment. And because every locale is unique, there cannot be a blueprint used by investors across the board. Everyone has to be innovative while still being logical, in order to ensure the interwoven systems exist in harmony. In summary, when it comes to seeking solutions to system problems, critical thinking does not put restrictions on how far you can go. What matters is that you will have identified current problems, or those that might potentially mar the relationship between your system and others. Sometimes the solutions you find most suitable may not be the most popular, but sooner or later, when the system is working like clockwork, stakeholders will be glad you adopted the most appropriate measures.

When considering the role of critical thinking in the context of systems thinking, take into account the varying needs of different systems—systems of the human body, which include the circulatory system, the renal, the digestive, the lymphatic, the muscle, the nervous and the rest of the systems; social systems, such as those of clans and other small communities, which the famous American sociologist, Talcott Parsons, described as particularistic ascriptive; religious systems, which the sociologist termed particularistic; economic systems, which include the market economy, the traditional or subsistence economy and their constituent systems and the many other systems that influence people's daily lives.

The important role of critical thinking in systems thinking can be summed up in a couple of points:

1. It helps to sharpen thinking skills in that you have to learn where things are going right with the system as well as where they are going amiss.
2. You have, of necessity, to think of the most suitable way to alleviate problems, taking into account the factors at play, including resources and effects on other systems, and discerning ways of strengthening the things that are going well.

Critical thinking skills are thus beneficial in all fields including finance, management and others.

Critical thinking is also important in the modern world where knowledge has gone global. With advanced communication technology, a discovery made in one corner of the world is known to the rest of the world in minutes. Likewise, misinformation can also be disseminated just as fast. It is therefore imperative that anyone in possession of critical information be capable of critical thinking so as to know how and when to disseminate it and to whom. At the same time, as a critical thinker yourself, you need to know what information to utilize and for what purpose. It is safe to say that, just as there is a global economic system, there is also a global knowledge system, and the same care accorded the economic system needs to be accorded to the knowledge system because both are interrelated.

You can understand this even better if you think of the information system releasing credible data on the susceptibility of certain geographical regions to terrorism. Do you think people will rush to invest in vulnerable areas? In releasing credible sensitive information, the information system will effectively influence the distribution of economic resources and the global economy at large.

As a critical thinker, it is also important that you wisely choose what information to embrace and what to cast aside, whom to release your information to and the timing, because the information system does not hold information for its own sake but for use by other systems, some of which may use it in unscrupulous ways. Even when there is no risk of information being misused, poorly timed release of information may cause your company losses or missed business opportunities, whereas suitable timing can put your business on the national or international map and raise your fortunes.

In fact, when it comes to matters of information and knowledge, it is imperative that you know how to identify information that is viable and how to apply it, because the body of available information is simply impossible to handle. Without the skills to sift information so that you are left with only what is useful to the situation you are dealing with, you can easily become overwhelmed and feel unable to proceed. At the same time, there is good information that can become deadly if released to the wrong people.

A critical thinker does not hold or release information merely for its sake, but rather analyzes it and evaluates it in the right context to see how helpful it can be. It is like having numerous natural spices in the kitchen. Whereas they are all great for health, each of them has its appropriate culinary use. For example, red pepper has a nice flavor and is nutritious, but if you put it in tea or coffee you are probably not going to make your visitors happy. Likewise, salt goes a long way toward making food edible, but too much salt can affect a person's health system.

Overall, when it comes to suitable application of information, critical thinking is important.

Main System Areas Enhanced by Critical Thinking

(1) Enhances communication

When you are good at critical thinking, you are able to express your ideas in a manner that is easy for other people to understand. That is fundamentally because, before seeking to express those ideas to other people, you will have analyzed and appreciated them yourself. When you are communicating ideas or presenting arguments whose content and context you well understand, you do it well. And, obviously, communication plays a key role in the success of systems thinking.

If you take the example of the new company being established in a quiet rural location, learn about the history of the area, the culture of the people and their outlook towards visitors. Based on this information, analyze how your company activities are likely to affect the lives of the people and weigh the negatives against the positives. You will then need to communicate effectively with local leadership and the residents themselves in order to take them through your thought process. Unless you can communicate effectively to show them that your intentions are good, not just for business but also for them, you may encounter a hostile reception that will, obviously, not be good for business.

(2) Enhances innovativeness

When you analyze a situation and possible ways of improving it, you will find yourself evaluating certain ideas under varying contexts, in order to see which ideas can help diminish existing challenges or enhance current strong points. You can only do this if your mind is imaginative, because you will not be directly implementing the ideas at this juncture but rather visualizing how they will work. It is at this stage that you are able to see ideas that are farfetched or misplaced, and ones that have potential to improve your systems. If critical thinking is not involved, one just collects facts and places them on the table, but as long as critical thinking is concerned, your creative mind must be engaged.

(3) Enables self-evaluation

Critical thinking ensures individuals, especially management staff, do not change policies and mode of operation haphazardly. When changes are made at random, the rest of the staff and everyone else concerned can feel left out and ignored, and the feeling can cause apathy at the workplace, at home, in the community, etc. On the contrary, when changes are effected only after critical thinking, there is room to lay out the pros and cons of maintaining the status quo vis-à-vis implementing particular changes. It is easy for stakeholders to back you if you explain the situation to them as backed with credible information.

In short, the fact that you cannot just wake up one morning and change the systems in your home, where you live with your wife and children, or your business, without any explanation, challenges you to be responsible and to always self-reflect.

(4) Establishes a scientific foundation

Whenever you want to present an argument of a scientific nature, of necessity, you have to support it with facts, showing what your observations are and how different actions involved relate to one another. Even when you have a strong gut feeling, you are called upon to do research and find supporting evidence before your argument can be accepted. All that data collection through observation, analysis, evaluation and deduction is part of critical thinking.

(5) Advances democracy

One might wonder what democracy has to do with critical thinking, but considering that democratic and undemocratic processes affect people's daily lives, critical thinking cannot be ignored. In undemocratic processes, management succeeds either through dictatorship, where people who are led are forced to play sycophancy, or through brutal authoritarianism. On the contrary, democracy thrives where critical thinking is involved because situations are evaluated and judged on merit, and recommendations are made on the basis of facts and supported arguments.

Still, even where people are free to make their own decisions, it is imperative that they embrace and practice critical thinking. Otherwise they may be influenced by lobbyists to accept policies that are not favorable to them. In communities where people think critically, they are able to take note of promises made by politicians, evaluate them and then distinguish those that are doable and beneficial from those that are mere vote-bait.

(6) Enhances a person's ability to think

Critical thinking is very helpful in advancing a person's capacity to think. There are people who easily become overwhelmed by a problem and cannot think of potential solutions. However, once you are adept at critical thinking you can quickly evaluate a situation and get a general idea of what the solution will be.

Just analyzing a problematic situation requires thinking. So the more you engage in critical thinking, the more your mind is prepared to think with clarity. You also will build stamina and focus when you are used to critical thinking, as opposed to some people who find it very difficult to concentrate on an issue beyond a couple of minutes.

(7) Helps navigate the global knowledge economy

For many years, people spoke of the money economy with admiration, but today the greatest currency is knowledge. Advancement in communication technology has made it all the more important for individuals and organizations to be able to sift information, because there is massive data landing on people's reading desks and being transmitted on electronic and print media, not to mention the overwhelming mass of data on the web.

People who have critical thinking skills know that you do not rely on any information to make an important decision unless you have analyzed the situation that requires attention and have identified the core of the problem—the same things advocated in systems thinking. Even when it comes to finding an appropriate solution to the problem, you only identify a solution if its application is going to solve the problem at hand without causing new ones. The same thought process applies to systems thinking.

It's crucial to use information well, as opposed to misusing it. Information fit for one system may not be suitable for another, and a critical thinker is able to tell exactly where to apply specific types of information. In short, critical thinking enhances proper application of information.

Critical Thinking Not Yet in Education

Many people who have heard of systems thinking and the good results it produces in organizations may not know how to apply it in their own different systems. Some may think the approach works well only in companies because they fail to understand that some systems or subsystems, or even parts of the system, are not physical in nature.

Often, systems are not tangible. However, after reading systems thinking-related content such as the one contained in this book, it becomes clear that systems thinking can be applied to any kind of system. Elsewhere in this book, examples have been cited where systems thinking worked well even within social systems. This chapter will explain what systems thinking means within the education system. Other sections of the book have shown that systems thinking can work in the sector because it has been tested through pilot projects, but this chapter will help you see the importance of systems thinking from a different vantage point.

Systems thinking, being a management discipline, seeks to streamline the way different functions are carried out in relation to others and to enhance the relationships that already exist among various systems and subsystems. Systems thinking has a lot to do with interactions, and that makes it very suitable for the education system where interactions between different stakeholders pose a challenge to modernization of the sector because of their diminished effectiveness.

For the education sector to perform well, the elements expected to interact effectively include learners or students, teachers, administrators, digital material, learning targets and others. Systems thinking can be used to enhance efficiency in the performance of the system that comprises those many elements. The approach makes use of data to construct a useful system meant to facilitate efficiency and usefulness for all.

There are instances where educational stakeholders attempted to adopt systems thinking and it did not seem to work well. Confusion sometimes results when schools try to digitize their systems due to overlap and glaring inefficiencies particularly where software and content are concerned.

Another challenge is piecemeal implementation of systems thinking. This patchwork approach is ineffective. Of course, there are instances of individual teachers doing an exemplary job just by sheer ingenuity, but overall, the education system in the US has not yet managed to reap the benefits of systems thinking.

There is, thus, dire need for systems leadership in the educational sector. This means people who can assess the situation critically, involve relevant external systems in the discussion and commit to adopting systems thinking in a holistic manner, rather than piecemeal or patchwork.

Chapter 4: How to Influence Matters Through Critical Thinking

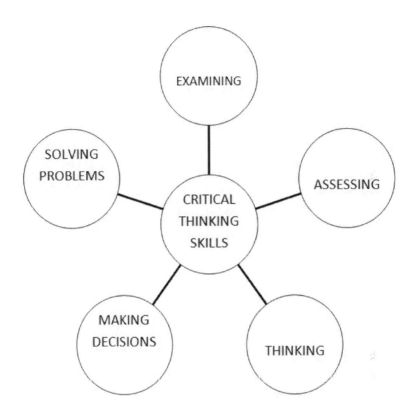

One of the important things thinking in systems accomplishes is the ability to influence the way things happen in various systems, and by extension the way things affect us. Through critical thinking, for instance, you are able to appreciate any prejudices you have and alleviate them through a thorough analysis and acknowledgment of situations and systems. According to one French philosopher, Jean De La Bruyere, life happens to be a tragedy for those people who are influenced by feelings, but a comedy to people who are influenced by thoughts. At the same time, some cognitive psychologists believe that your feelings are a direct result of your thoughts.

Whatever expert viewpoint you favor, it is undeniable that it is good to be able to tailor your thoughts to match favorable emotions. This is tantamount to saying that, through critical thinking, you have the power to look at a situation beyond its face value and analyze it from different perspectives so that you end up making deductions that offer hope to everyone involved. From this premise you can see that emotions and thinking are not mutually exclusive but rather go hand in hand. It is also safe to say that emotions do not necessarily precede thinking.

It is therefore liberating to know that when communities clash because of differences of opinion, possibly owing to variances in cultural orientation, it is possible to re-orient those communities through critical thinking and to help them see things from different perspectives that are favorable to peaceful coexistence. The same principle of working on one's thought process, with a view to producing positive emotions, can be applied at the workplace and in other systems as well.

What is important to understand at this point is that critical thinking can guide your thinking, and that can in turn help you appropriately control your emotions. Critical thinking can also help you adopt the right attitudes towards other people, situations and other systems in general. This means you can change your outlook to issues, courtesy of critical thinking, and be able to look at issues from a carefully chosen perspective as opposed to remaining captive to stereotypes and other culturally-ingrained views. You can also develop the capacity to guide your thought process in a way that helps you keep your mood balanced.

Critical thinking is beneficial even to the smallest of systems like that of oneself, because when you are able to control your feelings, you avoid taking actions that are detrimental to yourself, your neighbors, your family, institutions you are associated with and so on. As an individual, the power you get through critical thinking helps you avoid entering into stressful situations. Instead, you will seek out situations that are conducive to your overall health. On the contrary, people who have no critical thinking skills are likely to move through life without much control, influenced by the elements they encounter along the way. This means that if they are entangled with systems that are harmful to them, they will fall victim, unlike critical thinkers who evaluate matters, make deliberate decisions to effect changes that are beneficial or change course.

How to Tame Emotions Through Critical Thinking

It might seem challenging to think of how you can set aside feelings and engage reasoning when there is an emotional issue at hand, but it is doable. You only need to learn and practice the skills of critical thinking. Remember, in critical thinking you are able to neutralize your emotions by engaging your cognitive skills. That is why you will notice that critical thinkers are unlikely to act emotionally when something disruptive has happened. They have trained themselves to tame any feelings boiling within them, and to instantly engage their critical thinking skills.

Critical thinkers have feelings like everyone else, but they have practiced the art of taming them and looking at the core of the issue at hand. After analyzing and evaluating the issue, a critical thinker always seeks solutions to the problem. To ensure you are effectively succeeding in critical thinking, it is good to have some pointers that can help you assess your behavior.

How to Know You Are Engaging in Critical Thinking

- The points you put forth can be supported by reasoned arguments.

- You are evaluating a situation from different perspectives and not only one or two that you favor.

- You are open to different interpretations of the situation, meaning you are keeping an open mind.

- You are receptive to new evidence, explanations or even fresh findings.

- You are open to reassessing the information already in your possession.

- You are able to set aside preconceived ideas, biases and prejudices.

- You are ready to accept the existence of varying possibilities that are within reasonable parameters.

- You are not prone to jumping to hasty conclusions.

Important Steps in Critical Thinking

Whereas there are some people who are naturally patient and logical and generally have critical thinking skills, other people either glean skills from other knowledgeable people, or they read books like this one and then practice the skills they learn. Whatever way you use to acquire the skills relevant to critical thinking, it is important to constantly practice so that you master them, and in due course they will become part of you.

Thankfully, you do not have to grope in the dark in search of crucial critical thinking skills, as there is a well-defined path you can follow to become a master. Remember, the skills alluded to here are those that you can use to help you solve a problem.

(1) Pinpoint the problem

It is important to identify the actual problem in a system because without this you can spend unnecessary resources trying to alleviate a nonexistent problem or addressing the wrong systems area. Sometimes, when there is chaos, people correctly notice that something is amiss, but there are times when that chaos can be sorted out by simply improving communication. Such communication may require just a meeting or two between parties from different affected systems and no other resources whatsoever.

Yet, if you hastily delve into problem-solving, without sparing some time to identify the real problem, you might end up unsuccessfully spending money buying extra material, engaging new personnel and making all sorts of investments in a bid to eliminate the chaos. Just the sight of chaos triggers myriad suggestions on probable cause, yet your guiding factor should be the gist of the real problem.

If there is a real problem plaguing the system and you are able to single it out, you can proceed to analyze it in a systematic manner. In the meantime, you need to be aware that an aspect of the system can harbor the problem while still being able to contribute positively to the overall welfare of the system. How do you then proceed?

As a critical thinker, you need to analyze the role played by the affected factor and weigh the advantages it brings to the system against the disadvantages. Then you can confidently make an informed decision to let the status quo remain, to make minor changes or to overhaul the affected area. As one clinical psychologist, Barry Lubetkin, has mentioned, intelligent people are able to pinpoint the problem, weigh prevailing advantages against the disadvantages in a systematic manner and then articulate that problem in a way that is comprehensible.

(2) Analyze that problem

The problem has been identified, all right, but you need to reassure yourself that it is real and not imagined. That is one reason you need to analyze the situation with an open mind. Remember, if this is not your personal problem you will need to convince other people within your institution, or even other parties, that you may be directly or indirectly affected by the presumed problem. Needless to say, you need to be convinced of the problem yourself if you are going to bring other people on board, and that requires that you not only be well informed about the issue, but also have tact.

In your analysis, ensure you do not miss anything significant. Try and analyze the issue from as many different angles as possible, and then determine if the problem is really solvable. There are problems that can be so difficult to solve that you recommend a system replacement, depending on what that system is. Companies that do very well, for example, take time to find out if their machinery requires constant maintenance or needs to be written off and replaced. On the contrary, some companies that do not engage critical thinking keep on maintaining machinery as it breaks down, and at the end of the year the cost of maintenance will have eaten too much into the company profits.

Another aspect of good problem analysis is looking at the problem broadly as opposed to adopting a narrow perspective. In the former situation, everyone involved is able to see what the true position is and to easily appreciate the size and seriousness of the problem. Only after such appreciation can you determine if you are in a position to solve the problem, and if so, whether you require extra help from somewhere.

(3) Consider more than one possible solution

Whether you consider yourself capable of solving the problem alone or need to hold a brainstorming session with other people, it is important to have more than one solution to compare and consider. As the adage goes, two heads are better than one. Different minds can bring together different perspectives to help you understand the problem in new light or even to appreciate it better.

In a certain department you may have a problem of abrupt absenteeism, for instance, and, when consulting with present members of the department, float the idea of enlisting the services of temporary employees as a solution. Whereas you might expect this to be an offer met with relief, helping employees feel less overworked, they may inform you that every time they have engaged temporary workers there are problems such as broken equipment. Obviously, the reason you wanted to engage additional hands was to save the company production losses, and since maintenance costs are just as bad, you are likely to shelve your idea about temp workers.

(4) Check whether your suggested solution is suitable

It is possible to identify a solution among many which is cost effective and easy to implement, but if it is alien to the people in charge of implementing it you still have a problem in your hands. As the same time, a solution can be good, and the people charged with its implementation might like it, but you need to find out how it affects other systems.

Where relationships between systems are concerned, what you should be looking for is a solution that is not only good for your system, but also acceptable to other affected systems. Keep in mind that there is no one solution that can alleviate all problems, and that is why comparisons are important so that you can pick out a solution that promises to affect other systems the least negatively. Always consider the prevailing circumstances before selecting what appears to be "the best" solution.

(5) Take firm action

Firm action is an action that you confidently stand by. You can only claim to have completed the critical thinking process after you have taking the action necessary to solve the problem. If, after your critical analysis, you realize there is no solution that can improve the situation under the prevailing circumstances, or that the most fruitful solution is too expensive for you or for your institution, you can always say as much and recommend that the situation be left unchanged for the time being. If you make such recommendations, clearly stating your reasons, you will be considered to have taken necessary action and will be understood to have completed your critical thinking process.

The point here is that it is not enough to have wonderful analytical skills or the intelligence to think critically. You also need to be confident enough to take the bold step of implementing the appropriate action. One thing experts have observed is that people become more self-confident as they continue to practice critical thinking, and their sense of self-worth tends to become elevated. It has also been observed that, as people become adept at critical thinking, they become less and less afraid of challenges. This is very likely because they are confident of their ability to analyze situations and to seek and implement solutions to any existing problems, irrespective of what the best solutions are.

Overall, you will become better at problem-solving and feel more confident about your problem-solving abilities if you practice critical thinking on a consistent basis. You are bound to have a broader outlook to issues and to be in a good position to help others in the process of critical thinking. Of course, when solutions are sought through critical thinking, only the best are implemented, a move that is great for the system and other interrelated systems. It is important to note that the benefits of critical thinking transcend professional fields, political issues, businesses and social issues. As long as the solutions being sought affect systems, critical thinking is important.

How to Use Questions to Enhance Critical Thinking

In the normal course of things, we ask questions because there are things we would like to learn. When there is a problem to solve, often, questions help extract the right information from our sources. Without questions, people can bombard you with massive amounts of information, much of which is likely to be irrelevant, and, as already mentioned before, excessive information can be overwhelming. Even when you are trying to get information from books, magazines or the internet, you need to have a guide to ensure you only collect information that you need. Within the realm of critical thinking, that guide comes in the form of questions.

In fact, without the use of relevant questions, not only do you risk amassing too much information, but you also risk causing yourself confusion. In short, you need to learn how to design questions that are helpful in critical thinking; those that help you understand the problem and possible solutions in a logical manner. These questions guide your thinking so that it is relevant to the challenge at hand.

The purpose of questions in critical thinking is to set the agenda for your research and analysis. Whatever questions you pose help to tailor your line of thought, ensuring your thoughts are logical and cohesive. As a result, the information you receive in response to the questions is relevant to the situation you are handling. In order to trigger and enhance critical thinking, you need to design your questions in a certain way.

The mode of questioning we recommend in this book is one introduced by Benjamin Bloom, a renowned educator, popularly referred to as Bloom's Taxonomy. When you follow this mode of questioning, your questions are not one-dimensional, the kind that demand a yes or no answer, or such brief answers that do not help you with the way forward or provide helpful explanations. You also get to plan your questions in advance so that you consciously create questions that activate higher-level thinking.

One reason questions in critical thinking need to be carefully designed is that you want them to help you extract helpful information from your sources and make you knowledgeable about the situation. You want them to help you comprehend both the nature of the problem and the possible and feasible solutions. You also want the questions to help you apply the knowledge you gain to the situation at hand as you analyze, evaluate and synthesize it.

In reference to being knowledgeable, we are referring to the facts you gather and manage to remember, the opinions you welcome from various sources and other important concepts. When it comes to comprehending the problem and potential solutions, the focus is mainly on your ability to appropriately interpret the information gathered. Questions should help you correctly interpret information and use it appropriately to help the situation.

The questions you design should lead you to understand the situation so well that, in the future, if you are confronted with such a situation you will be able to handle it with ease. The questions help to enhance your understanding of the dynamics of prevailing internal relationships, which is especially important considering how interrelated systems are. Questions related to evaluation help to enhance your judgment of the situation so that you are reasonable as you set standards and decision-making criteria. Those questions designed purposely to help you in synthesizing information are meant to help you consolidate and organize facts in a cohesive manner so that you can easily visualize a clear picture of the expected outcome.

What to Seek as You Design Critical Thinking Questions

1) To gather knowledge

In this case you want to design questions that will lead you to gain helpful knowledge that is also relevant to the situation. To succeed in this, you need to first of all be aware of what you really want to achieve. Begin by recapping the things you already know about the situation or problem at hand. The way to do this is by reiterating facts, main ideas and relevant terminologies, as well as relevant responses already received from various sources. This is how you pose questions associated with knowledge:

- What, exactly, is this?

- How do you identify that?

- Why, as far as you know, did this happen?

- What, in your opinion, is the best way to describe this scenario?

- Exactly when did that happen?

2) To enhance comprehension

The kinds of questions you ask here are aimed at enhancing your understanding of the situation, and at the end of the day you need to be able to demonstrate that you understand the relevant facts and specific concepts provided. To succeed in understanding, it is imperative that you aptly organize the facts and concepts available, compare them and translate and interpret what they mean. As you do this, you need to allocate appropriate descriptions and clearly state the ones you consider to be most important. This is how you design questions meant to enhance comprehension:

- How, in your view, does this compare to that? How would you say this and that contrast?

- Can you give an explanation for this in your own words?

- In your view, what facts or ideas support this position?

- Do you have evidence to support this? If so, what is it?

3) To master application

The reason you want to master application of the knowledge at your disposal is that the problem you identified still persists, and if a solution has been tried before, then it has definitely failed. In short, at this juncture, when you have the knowledge you need and a good understanding of the problem at hand, the only thing remaining is good implementation of the solution. The relevant facts and information that you have should form the basis for your knowledge, and then you can choose a technique uniquely tailored to your situation. If you keep following the beaten path, the problem will probably recur like it has done, or it will not go away in the first place.

How to frame questions related to application of knowledge:

- What is the best approach for …?

- What are some great examples of …?

- How would you display your understanding of …?

- What, in your view, would have transpired if …?

4) To guide analysis

Analyzing a problem is important because it helps you to choose the most appropriate information to apply and the best technique to follow in solving the problem. Usually, after collecting information and opinions, you end up with generalizations, and these are not exactly helpful when it comes to seeking solutions to the problem. You, therefore, need to break down the information you have in order to extract credible material that you can use to support your arguments. It is only after you have broken down the generalizations into specifics that you can pinpoint any inferences or motives contained therein, and you may even be able to see problem causes very clearly.

This is how your analysis questions should look:

- Going by this piece of information, what inferences can you draw?

- How would you categorize that or that other?

- How would you classify that?

- Can you manage to identify the various sections of this part?

5) To direct evaluation

Evaluation here refers to your assessment of work done. Is your solution worthwhile? Are your ideas meaningful? Remember, all this is happening within the realm of critical thinking, and so your arguments and proposals need to be solid and able to withstand testing. Of course, it is expected that you have an established criterion or set of criteria against which your work will be assessed or which you will use to validate the ideas you propose.

This is how good evaluation questions are designed:

- How would you compare this and that?

- Which would you say is better between this one and that one?

- If I were to seek your opinion, what recommendation would you give me?

- How do you feel about Y's participation compared to Z's?

- In your assessment, what was the significance of using this or that in this context?

6) To enhance synthesis

At this point, you have not only assembled the relevant information, but you have also visualized what the situation will be like after your solution has been implemented. Nevertheless, in critical thinking you are always encouraged to prepare alternative solutions whenever possible, just in case the first one you try out hits a snag. After all, as has already been noted, a solution can be very good for one system but fail the test when it comes to how it affects other systems.

What you are seeking now are ways to consolidate the already sifted information and to try to see how it can be applied differently from the way it has been used before.

The best way to design questions related to synthesis:

- Supposing you had this or that happening ...

- Can you imagine any way we can circumvent this particular problem?

- Do you have an alternative interpretation to this?

Chapter 5: Major Strategies to Enhance Critical Thinking

If you think about what critical thinking entails, including understanding a problem, gathering and sifting information, organizing and analyzing it, you will see how progressive it is. The entire process of critical thinking makes your mind sharper, irrespective of whether or not you find the ultimate solution to the challenging situation.

Ordinarily, people prefer to take the easiest option in life, and that entails moving with the flow as opposed to establishing why things are not working efficiently and seeking solutions. However, anyone who wants to excel in life needs to master critical thinking because that is the only way they can unleash their potential in whatever field they are in.

Systems can only work efficiently on a continuous basis when the people running them are capable of critical thinking. As has been pointed out, people have to be prepared to change their thought processes when they are engaged in critical thinking, and by extension, when thinking in systems. After all, nothing is just about oneself when it comes to thinking in systems, because whatever system you choose to work on, you are bound to find other systems embedded into it.

Essentially, therefore, when you think about problem-solving, you need to think in systems, and the critical thinking you do needs to be in the context of working within systems. However you choose to go about your problem-solving, you need to remember that no system operates in a vacuum or in isolation. Therefore, while being thorough, you also need to learn to be tactical, so that you do not destabilize the entire body of systems.

The strategy you choose to apply to enhance your critical thinking skills is important. In this chapter you are going to learn the best strategies to follow to ensure an efficient and fruitful critical thinking process. These strategies are meant to help you fundamentally become a critical thinker, and not just to give you skills to use in solving one type of problem.

- Endeavor to make connections relevant to the problem

In this effort, you need to think deeply about the problem, meaning what will be tested is your willingness and capability to direct your thought process towards the relevant issue. Critical thinking calls for a conscious effort to focus on the chosen issue; otherwise the mind might wander and not form any helpful thought process. According to experts, unless you give your mind a deliberate lead, it keeps switching from one focus to another every ten seconds.

- Formulate high-quality questions

The reason you need to design high-quality questions to help your critical thinking process is that they provoke clarifications and other helpful responses. It will not be helpful to create casual questions if you have already determined there is a problem that requires attention. To invite helpful answers, it is important that you tailor the questions in a style that leads the other person to focus on the problem and information related to it. In order to create appropriate questions to help in critical thinking, you need to know what to seek and where to focus.

Foundation for Creating High-Quality Questions

Use evidence as well as logic to support your thinking.

- Make sure you analyze, reason and evaluate the situation.
- Ensure you interpret any information you have beyond face value.
- Utilize a blend of ideas from a wide range as opposed to confining yourself to one or two of ideas.

- Seek answers to complex questions but ensure they're relevant to the problem.
- Think through any decisions before you make them.
- Generate varying options and evaluate each of them before you can settle on one in your attempt to solve the problem.
- Always make use of details so that you can end up with the correct meaning of the information you receive.
- Engage high-level thinking when designing questions to solve real-life problems.
- Make critical thinking part of you so that the questions you ask are meaningful.
- Use preset criteria to determine how valuable your ideas and proposed solutions are.
- Ask questions that show your reflective nature.
- Take steps that are bound to lead to problem-solving.
- Proactively question the credibility, veracity and relevance of any information you plan on using whether you have received that information from observing, listening or reading.
- Be well informed.
- Be open to looking at the situation from different perspectives.
- Try to find solutions that are not only fresh, but also better than what you had before.
- Be prepared to explore additional possible solutions.
- Whatever different points of view are presented to you, be prepared to examine and assess them.

- Value and respect ideas suggested by other people.
- Before implementing any ideas or taking any action, assess the potential consequences.
- Even though you are encouraged to consider other people's ideas during the critical thinking process, you also need to make a point of making independent evaluations.
- Pay attention and adhere to fundamental universal intellectual standards.

What are those basic standards? They include standards of clarity, precision, accuracy, relevance, depth, breadth and logic, as well as significance. For you to be adept at critical thinking, you need to practice it continuously and not only when there is a system failure or other complex problem to solve. According to experts, it helps if you pick one of the mentioned standards and devote one week to practicing it and internalizing it. How do you do that?

The best way to maintain a particular intellectual standard is by continually being critical of yourself, not with the intention of making you feel deficient, but to make you realize when you are not performing up to par. Take the universal standard of clarity. If that is what you have chosen to practice this particular week, don't let yourself get away with being vague in your communication. When you have consciously chosen to internalize the standard, chances are you will realize whenever you have communicated in an unclear manner, and will instantly seek to correct yourself. Similarly, you will be quick to notice when other people are not communicating clearly.

Keep in mind that the same way you need other people to communicate clearly to you is the same way other systems need to be communicated to with clarity. Unless communication among systems is clear, it cannot be effective, and consequently there may be misinformation and inefficiencies in the way processes work.

Miscommunication is one reason institutions perform poorly or incur heavy losses, even when they have invested heavily in resources.

As you continue to practice communicating with clarity, you will soon become a master in communicating explicitly, knowing when to elaborate issues using different words and helpful examples. You are also likely to be good at using personal experiences to drive your point home and will use appropriate figures of speech or even analogies.

- Endeavor to reshape your personal character

The idea of reshaping your character is associated with enhancing the intellectual traits you have. Those intellectual traits include humility, empathy, autonomy and perseverance, as well as courage. People who are genuine critical thinkers are not braggarts, and so they are unlikely to cause discomfort to other systems as they try to solve problems within their own systems. All the other traits of a critical thinker are also likeable, and so they often encourage other people to provide great feedback that is helpful in the critical thinking process.

According to experts, you can easily succeed in reshaping your character if you choose to practice one intellectual trait throughout one month. During that period, you are encouraged to be aware of the instances when you practice the chosen trait well, and also when you fail to live up to that trait. That way you will internalize the trait slowly, and it will soon become part of your character.

For example, in the month that you choose to practice humility, ensure you take note of the times you acknowledge being wrong, because such an admission is a mark of humility. At the same time, if you realize there was an incident when you denied the reality despite facts speaking for themselves, identify that as a failure of humility on your part. Doing that is likely to caution you against making a repeat mistake later, and more so, it shows you acknowledge markers of humility.

Along the same lines, you need to acknowledge instances when you have become defensive instead of appreciating divergent views and admitting your shortcomings when they are pointed out. When you consciously practice humility and admit failure when you do, the virtue slowly becomes ingrained in your system, and humility will no longer be something to consider but something that manifests in your everyday life.

- Adopt an accommodating attitude

In systems thinking, and particularly in critical thinking, it is not only your opinion that matters but also the opinions of other people who have an interest in the situation, system or systems linked to it. As such, much as messages and concepts may seem self-explanatory, it helps to listen to the meaning adopted by other parties. Even when it comes to a given situation, there may be other people looking at it from perspectives different from yours, and you need to give their perspective a chance.

It is important that you avoid taking situations and everything else at face value, but instead be open to other meanings and other possible implications. Any one situation is likely to have as many perspectives as the people interested in it, and in the context of systems thinking, there are likely to be as many standpoints as there are constituent or affected systems. When you develop an accommodating attitude, you take note of suggestions presented to you by different people, and then you use your skills to critically analyze each one of them as you evaluate the facts at your disposal. Then you make a point of gathering even more relevant facts in order to improve your understanding of the perspective given.

After taking the different suggestions and perspectives into consideration in this manner, you are better placed to make a decision on how best to proceed in the problem-solving process.

- Acknowledge group influences on your character

It is common knowledge that human beings are fundamentally social animals, and so it is inevitable for individuals within a social setup to influence one another. There is also behavior that people within a group adopt only because they are members of that group, and which they would not dare adopt if they were not part of the group. Group psychology and influence is real, and you need to shed it if you are to become successful in critical thinking.

Critical thinking enhances your personal development as an individual, and because of the decision-making freedom you gain from it, you are bound to become a happier person. Of course, there is some good that comes with socializing and being part of certain social groups, but those circumstances become a bad influence when your bond with these groups inhibits your mental independence as well as your critical thinking prowess. In short, enjoy whatever benefits come with being part of your social groups, but be aware of the risk of losing the essence of who you are, including your valuable qualities as a critical thinker.

- Acknowledge your emotions

The fact that you are a critical thinker does not mean you are not vulnerable to emotions that can easily mar your judgment. However, when you are aware you have strong emotions regarding a particular situation you are trying to address, you can consciously try and push them to the periphery to enable you work objectively.

After acknowledging the existence of those emotions, try and establish their basis and proceed to find alternative ways of looking at the situation. By acknowledging your strong emotions and reining them in, you will find they have a diminishing negative influence on your ability to work efficiently. In short, when you are adept at critical thinking, you are better able to control your emotions, which means you are more patient, accommodating, understanding and generally better placed to think in systems.

Chapter 6: Critical Skills in Systems Thinking

For many years there have been myriad challenges facing the world. Some of these challenges include drug and substance abuse, unemployment, income distribution, environmental threats and AIDS. We have made little progress in solving these challenges. In other cases, the problems have only gotten worse and new challenges have emerged.

One might wonder why, despite technological advancements, the world has made little progress in solving such societal problems, and the answer lies in how we think, how we learn and how we communicate. We continue to use outdated methods in identifying problems, and very often what we pinpoint as the problem is not actually the core of the issue. In fact, it may be a problem on the periphery, and when everyone focuses on solving it the main problem is meanwhile messing up the system.

Other times what we see as the problem, using outdated methods, is only a symptom, and unless we recognize that and embark on the search for the real problem, many of the problems will persist. In fact, in the process of pursuing the wrong target and taking action on it, we end up creating more problems for the system and those associated with it. However, in systems thinking, we believe that we can change how we learn and how we think, as well as how we communicate. This is what has been described elsewhere in the book as changing our thought process, which then changes how we behave or act.

As we pursue this approach, embodied in systems thinking, we are able to face and offer solutions to real, persistent problems, and others that continue to emerge. Most importantly, in order to achieve the best results in education, we need to evaluate what and how students are taught.

Here we shall begin by defining learning, thinking and communicating, as it will help us vividly see the skills that need changing. Understanding these concepts will also make clear the current situations that are responsible for hindering the preferred process of change, which is systems thinking, and what makes systems thinking so important in this process of change.

Thinking

Thinking can be defined as having a thought or being able to reason or reflect on something. Thinking involves creating mental models and then acting on them so as to make decisions and conclusions.

A mental model is a selective concept that we create from reality. We simulate these models so that they can produce meaning in whatever situation we are in, and then they end up guiding us in taking appropriate actions and making suitable decisions.

For example, when you think about people, it does not mean that those people are right in front of you. The reality is that they are only in your mind in abstract form. Yet you are able to choose certain aspects of that person that relate to whatever you are thinking about them. If you seek to understand why your child's school results are deteriorating, you will probably ignore the color of her hair in your mental model, because it has no relationship to the concept you want to pursue. However, when selecting a birthday dress for the child, it's important to include the color of her hair in that mental model.

Mental models and others are simplifications of reality, and they sometimes omit some aspects of that reality. When making any rational decision, it is always important to have a mental model, otherwise you may take action out of intuition, a move that is not verifiable. In short, without a mental model it's difficult to explain the basis upon which you are taking a particular action.

Constructing Mental Models

When constructing any model in any field, be it mental, scientific, academic, environmental, family-related or even societal, some basic questions must be answered:

1) What components should and should not be included in the model?

2) How should those components be represented?

3) How should the relationship between the components be shown?

How to Select Components to be Included in the Model

Selecting the most appropriate model is determined by how widely and deeply you want to cover the specific topic. Systems thinking offers several skills that help you become adept at selecting what is best to include in their model. These skills are also called filtering skills because they help us filter out what is irrelevant to the mental model.

Systems-Thinking Filtering Skills

1) 10,000-meter thinking

   The 10,000-meter thinking skill was inspired by the view that one gets from an airliner on a clear sunny day. From high above you are able to see a very wide horizontal range of area, but very little of the area can be seen vertically.

2) Systems-as-cause thinking

   This thinking skill holds that mental models should only include those components that contribute to the phenomenon of interest. A model should not include any external forces. For example, if you are holding a slinky with both hands and you compress it, then remove the hand that is holding it from below, the slinky will oscillate. What is the cause of the oscillation?

   Most people will say the oscillation was caused by the removal of the hand, while others will attribute the oscillation to gravity. However, in systems thinking we attribute the oscillation to the slinky itself. In order to appreciate this answer, imagine if you did the same experiment using a cup. Would the result be the same? Would oscillation occur?

No oscillation would occur if you performed the same action on a cup. The oscillatory behavior is strictly within the slinky itself and does not involve any other item. However, this does not in any way mean that gravity and the hand are not important in the action of oscillation. It only means that they do not need to appear in a mental model that seeks to explain why a slinky oscillates.

3) Dynamic thinking

This is the third filtering skill. It is similar to 10,000-meter thinking only in that it emphasizes behavioral dimensions rather than structural dimensions. In most cases, we are caught up with the details of structures and events while we ignore the patterns that led to that structure or event.

History students, for example, are keen on memorizing the dates when great leaders were born, when battles started or ended, when agreements were made and so on. However, before the arrival of those dates, there were indisputably events that led to the buildup of those major events. For example, before the United States attained its independence from England on July 4, 1776, there were a lot of conflict and negotiations.

Dynamic thinking encourages us to push back and see the patterns which those specific events are part of. The consequence is that, through mental models, we are able to view reality as dynamic as opposed to being static.

If we compare the divide-and-conquer perspective and the systems-thinking approach in creating mental models, we find that divide-and-conquer gives us a stark mental model, that is, a vividly accurate picture of the reality. On the other hand, in systems thinking we are able to create a mental model with a broad perspective.

To solve the many challenges and problems that face us as a society, we need to have a broad mental model. This can only be achieved by altering our education system to focus more on horizontal thinking skills.

How to Represent the Components in the Model

After selecting the activities to include in the mental model, we now seek to view how we can represent those components. One major hindrance to students developing their skills in representation is that various academic fields have their own way of representing their content. The different ways include symbols and icons, different meaning of terms and also guidelines on how to represent information in specific fields. Students struggle to memorize the subject matter, specific vocabulary and to grasp methods of representing information in that specific field. This leads to students acquiring subject-specific skills. As you may imagine, narrowing a person's skills to one field limits their ability to comprehend other areas and how they affect issues outside a narrow area of knowledge.

In systems thinking, we use the language of stocks and flows. This language is derived from Esperanto, which is a lingua franca which supports cross-disciplinary thinking and therefore offers horizontal-perspective thinking. In mental models that have been formed using the language of stocks and flows, it is easy to clearly distinguish the components of the model. The things that accumulate are called stocks, while those that flow are called flows.

Stocks denote conditions in a system, while flows represent elements that cause those conditions to change. That is surely easy to comprehend no matter what your area of specialization is. For example, in a situation where we have water in the clouds, evaporation occurs. Then there is no more water in the clouds. Thus, water in the clouds is the stock. The process that changes that situation, evaporation, is the flow.

Let us consider a person whose body weight has continuously been increasing. The decision the person takes is to stop eating junk food, and also to enroll for an exercise program which he makes sure he adheres to. This person will have reduced the volume of the gaining flow but increased the volume of the losing flow via increased caloric expenditure. Does the person continue gaining weight or does he start losing weight?

There are two perspectives in answering this. One is that if the inflow remains higher than the outflow volume, then the person will continue to gain weight, albeit slowly. However, if the outflow surpasses the inflow then the person will start to lose weight.

How to Develop Content-Representation Skills

It is very important to distinguish between stocks and flows because failure to do so in mental models, which must infer dynamic patterns of behavior, may lead to drawing erroneous conclusions.

Since stocks and flows are content-independent, their use contributes to a student's general understanding of representation in whichever field they are applied, because they are not subject-specific.

It is important to note that, as the student participates in subject-specific representation skills, he or she also builds on general representation skills because the language used is content-transcendent and, hence, a virtuous cycle of learning is created. The cycle is effective in that all content areas benefit from the activities that are taking place in the other fields. Instead of knowledge from one field affecting the other, it reinforces the other fields as well. This is a clear example of closed-loop thinking, which is one of the systems-thinking skills.

In order to effectively communicate using the language of stocks and flows, students need to have a clear understanding of operational thinking. So as to build the students' ability in representation skills, it is important for them to be taught the language of stocks and flows and operational thinking from an early stage in the formal education process, probably in grades four or five. Most importantly, it will improve the students' ability to think broadly, with horizontal thinking skills. Fortunately, the curricula in these lower grades are still flexible enough to accommodate some of these changes.

How to Represent Relationships Between Components

The final question is how we should represent the relationships between these components that we included in our mental models. Here, we have to make certain assumptions on the generality of those relationships. Every society has a general agreement on how some of these relationships work. For example, in Western culture, the unspoken agreement is that reality works through a structure of serial cause-and- effect relationships. In fact, this has been, for the most part, the traditional approach to issues. In that approach, when something happens, the action directly leads to something else happening and so on and so forth. However, not all structures subscribe to this concept. In fact, systems thinking itself does not work in this manner of cause and effect.

Cause-and-effect thinking is also called laundry thinking or critical-success-factors thinking. This model is built on four assumptions which are content-transcendent. The assumptions are used when creating mental models to address various social and personal issues, and also in various fields such as literature and chemistry to show cause-and-effect relationships. Since we have always subscribed to these relationships, sometimes we do not even realize that we are using them. They have been so repeatedly used that they are no longer seen as assumptions, but as part of reality.

Using these assumptions in the laundry list while creating mental models often leads to erroneous conclusions when simulating mental models. We will discuss the four assumptions of laundry-list thinking and then see how systems thinking seeks to offer solutions to the shortcomings of each assumption. What causes students to succeed academically? That is the question that is going to guide us in discussing the four assumptions.

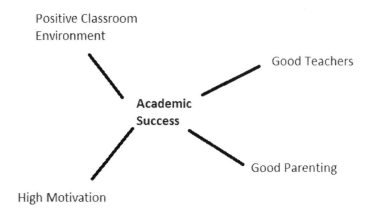

Most probably, your answer has a list that includes, among others, good teachers, good parenting, positive learning environment and motivation. The first assumption is that each causal factor operates independently of the effect. The story here can be told of good teachers, motivation and a positive learning environment causing academic success. Each factor is assumed to independently contribute to academic success .

This is certainly not the case, as in reality all the factors have to act together in partnership so as to achieve academic success. Good teachers ensure a positive learning environment and they also motivate their students. Motivated students, in turn, motivate the teachers, who ordinarily tend to become even better teachers, hence contributing to the academic success of the student. A positive learning environment creates teachable students, who, in turn, motivate the teacher to do even better. All this is made possible by having open communication between teachers, parents and students. The four factors in the laundry list model must work together in contributing to students' academic success. This creates a picture of an effect acting as a cause, rather than laundry-list thinking.

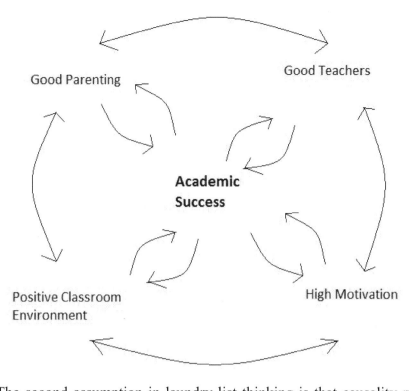

The second assumption in laundry-list thinking is that causality runs one way. In this case we move only from cause to effect:

Motivation ⟶ academic success

Good teachers ⟶ academic success

Positive learning environment ⟶ academic success

Good parenting ⟶ academic success

However, this is not true because the effect can also act as the cause as already alluded to above. For example, academic success also motivates teachers, students and parents to become even better. Academic success is a cause of the four factors just as the four factors are a cause of academic success.

Academic success ⟶ motivation

Academic success ⟶ good teachers

Academic success ————————→ positive learning environment

Academic success ————————→ good parenting

As we can clearly see, each factor affects another in the systems-thinking model, which is two-way or closed loop. The assumption is that causality runs one way and is static while the closed loop shows a continuous process in the dynamic view. The ability to view reality as being made up of feedback loops and to create mental models showing the relationship between these components is a systems-thinking skill referred to as closed-loop thinking.

Mastering this skill helps students to create more reliable mental models. In the social arena, solutions to problems facing society will not be seen as one-time fixes but rather as interrelated loops that will continue after the solution has been implemented. When we develop closed-loop thinking, we are able to anticipate consequences that we had not planned for. We are also able to identify the most efficient leverage points—the points at which a minor change may cause tremendous changes in the system. If these concepts are introduced in the earlier grades of learning, the solutions offered by future generations will, most likely, be more effective than those from a one-way-causality inspired generation.

The third assumption in laundry-list thinking is that the causal impact unfolds immediately without any delay. Feedback loops in both the natural system and reality occur in nonlinear patterns. Patterns such as these cannot occur in mental models that are linear.

The fourth assumption in laundry-list thinking is that impacts from causes are felt immediately. For example, we assume that a positive learning environment such as good ventilation, good lighting, adequate spacing and adequate resources will immediately boost the academic success of a student. However, this does not happen because of inevitable delays which are involved while the learning environment reaches appropriate standards.

It is important to include delays in mental models so as not to undermine the reliability of simulations from such models. It is also important to include delays in any curriculum that seeks to develop effective thinking skills as this helps students to know where to include them in their mental models.

Initially we started by stating that our education systems hinders the development of students' thinking, learning and communicating skills. We have seen how the education system restricts the selection and representation of components in mental models. We have also offered a systems-thinking approach to these restrictions, showing how to overcome them. If students practice the given skills, i.e., 10,000-meter thinking, system- as-cause thinking, dynamic thinking, operational thinking and closed-loop thinking, they will create more reliable mental models and bring forth realistic conclusions.

Learning

The first type of learning is called self-reflective, and we identified it in the critical thinking process. It occurs when a mental model changes its content or representation due to the outcome of its simulation. The second type of learning, which is inspired by communication, is called inspired learning. The level of learning here occurs depending on the quality of the feedback provided and the willingness to respond to the feedback. Sometimes we can also learn from the consequences of one's actions. This last method, however, may take time because the impact of one's action may also take time before they are fully realized.

It is important to note that learning, thinking and communicating are self-reinforcing processes, and unless there are changes made to the mental models, then learning does not occur. When we self-reflect on the outcomes of our simulations, it helps in changing our mental models, and thus, learning occurs. In cases where the simulation results are false, then self-reflective learning is affected.

It is therefore important that simulation results are realistic so that the learning process can be effective. In our current education systems, very little has been done to develop simulation skills. This means that we lose very critical feedback which is important in improving the quality of our mental models.

Communication

Communication here includes the feedback we get after analyzing a mental model and the outcomes we get. The current education system provides very few avenues for students to critique other students' mental models and their outcomes. In order to develop each other's learning through giving and receiving feedback, we need to be empathetic. Empathy is a critical value in systems thinking as it increases the value of true care beyond self. Being empathetic increases one's patience in listening to feedback and articulating it accordingly.

However, even with empathetic skills, we need a language that can be understood across various disciplines. This is where a content-based curriculum acts as a hindrance because, even if there was adequate time for students to critique each other's mental models and offer feedback, there would still be disciplinary segmentation which would hinder communication. Why is this?

Each discipline has its own jargons and symbols. This then makes it difficult for students outside that discipline to understand the content of the mental model well enough to be confident to share their feedback. This is where stock and flow language comes in handy. This language helps students to move comfortably from one discipline to another at will, with the capacity to understand what is being communicated in those disciplines. In this way, whatever is being learned in one discipline acts as a reinforcement to another discipline, rather than an impediment. In this way, too, students are able to have a horizontal view of the connections in the real world.

Chapter 7: Role of Emotional Intelligence in Critical Thinking

Critical thinking is essential if you are to succeed in thinking in systems, and for critical thinking to succeed, it is important to have a significant level of emotional intelligence. Emotional intelligence (EI), which is sometimes referred to as Emotional Quotient (EQ), is simply your capacity to recognize and acknowledge your emotions, evaluate how intense they are, assess if they are healthy or not and ultimately control them appropriately. EI is also seen in your ability to positively influence other people's emotions.

Think of a system like a company whose activities inevitably affect the neighborhood community. If it is a manufacturing company, noise from machinery is bound to fill the air—not a pleasant thing to the residents of the area. However, if you have people in management whose EI is high, they will have thought critically about the problem and devised ways of helping the community bear with the noise, embracing the company as an area development. Essentially, the company and the surrounding community are two systems that affect each other in their existence, and for both to succeed they have to be empathetic towards each other.

Sometimes the reason different systems are continually at loggerheads is that the people involved are not aware of how important it is to be emotionally intelligent. Before delving deeper into how best to enhance EI for the benefit of businesses, political organizations and all other systems, it is important to disabuse ourselves 0f certain notions associated with EI.

Being emotionally intelligent is not the same as always being:

- Agreeable

- Optimistic

- Happy

- Motivated

- Calm.

Although these traits are wonderful to have, they are not synonymous with EI. According to Dr. John Mayer, an American psychologist, EI is the ability a person has to use reason to tame emotions and to direct any emotional information he or she may have, in a bid to stabilize and improve his or her thought process. Once a person can control his or her thought process, irrespective of emotions, that ability to think critically and to enhance relationships among systems will remain strong.

This is a great achievement because, as has already been mentioned elsewhere in this book, thought process influences actual feelings. Of course, if a person does a critical analysis soberly, without being unnecessarily influenced by emotions, credible results are bound to occur that will eventually enhance efficiency within and among the systems involved. People with high EI become adept at critical thinking by virtue of their ability to control their emotions and to avoid stubbornly holding extreme positions.

In fact, EI people are better placed to discern other people's emotions just by being in close proximity to those people and observing them for a brief period. People with high EI are also able to sense the intensity of anger other people around them, and as a consequence they are able to detect when that anger has reached dangerous proportions. Likewise, they can tell when people's happiness motivates them to interact more socially and when their sadness causes them to stay aloof. In general, people with high EI are empathetic to others, and they are better placed to solve other people's emotional problems than people whose EI is just average.

Qualities of People with High Emotional Intelligence
1) They are able to relate amiably with other people

Emotionally intelligent people can moderate their own emotions and empathize with other people when they are emotional. Thus, they are able to relate well with others at the workplace, in their respective communities and elsewhere.

2) They are good at achieving their goals

For the same reason that they do not allow emotions, either their own or from other people, to affect them in a negative way, EI people often find it easier than others to achieve the goals they have set. This success is also enhanced by an emotional maturity that enables them to handle issues soberly even when the environment is emotionally charged.

3) They are good at nurturing healthy personal relationships

People with a high EI level are able to control their emotions and to balance them in a way that creates a healthy environment for developing and sustaining personal relationships.

4) They generally lead balanced lives

As has already been mentioned, people that have high EI are able to analyze issues soberly, and that helps them in their critical thinking. As a consequence, they not only enhance their problem-solving abilities, but also streamline their personal lives, critically managing their emotional trends.

5) They tend to be compassionate

People with high EI are generally compassionate, owing to their ability to empathize with other people.

6) They are generally happy

The reason people with high EI are generally happy is that they often do not harbor ill feelings towards other people. This capacity to shed such negative feelings is enhanced by their ability to be in touch with their emotions. After acknowledging any negative feelings creeping in, and after taking charge of them and controlling them appropriately, high EI people become happy, playful, creative and fun to be around.

7) They are great at resolving conflicts

Emotionally intelligent people are great at critical thinking, and their ability to enhance system relationships is exemplary. This should not be surprising considering their other qualities of being able to manage both their emotions and those of people around them.

How Modern CEOs are Succeeding Through EI

In today's business world, leaders have done away with traditional structures or organizations where management leads by commanding and controlling the workforce. They have realized that, for systems to work efficiently and businesses to become super profitable, operating through strict command chains is not necessarily most effective.

In fact, many successful CEOs appreciate how dynamic organizations are due to human involvement, acknowledging that there are both constructive as well as destructive energies at play at every moment. In traditional management methods, leaders seek to identify positive energies and try and harness them, whereas contemporary leaders have begun to learn to let organizations self-regulate through what is termed productive management.

For any system to self-regulate, EI is vital. This is because it brings out skills that create preferred behavior in people. In general, when an organization engages people with high EI, it becomes successful because of:

(1) Self-awareness

Emotionally intelligent people are self-aware, meaning they understand not only their emotions but also their personal tendencies and how they are likely to behave under different circumstances.

(2) Social awareness

Emotionally intelligent people can identify specific emotions in other people and discern their inner thoughts. People who are self-aware and have the ability to understand other people's emotions are able to nurture and safeguard relationships, something that is good for the organization as a whole.

(3) Self-management

In self-management, people with high EI are capable of being flexible in terms of adjusting to organizational needs as they arise. Also, since they are self-aware, they are able to control their behavior so that it remains constructive at all times.

Emotional intelligence should not be confused for soft skills, which are also good for management. Nevertheless, it enhances leaders' ability to make good decisions, accommodate stress, become good time managers, etc. Many contemporary CEOs who have adjusted to this modern way of management have taken their organizations to great heights of success and, in the process, earned respect among their peers and the rest of the world.

Since some people continue to wonder if EI or EQ is real or just hype, experts have found a way to evaluate it. Today it is possible to scientifically measure someone's EQ the same way Intelligence Quotient, or IQ, is measured. Incidentally, EI is not entirely new—people just take a little while to warm up to new ways of doing things. In his 1998 publication, *Working with Emotional Intelligence*, Daniel Goleman defined EI as a person's capacity to recognize his or her own feelings as well as other people's feelings, the ability to self-motivate and also to control personal emotions within relationships.

In fact, EI is not just good for business or for big systems; it is good for everyone within their different setups. In a religious context, Jesus encouraged Christians to love one another and to exercise compassion towards one another. Nobody translated that to mean using EI, but that is what it was about. Early philosophers were also known to encourage their followers to know themselves, a very strong aspect of EI. In psychology, Sigmund Freud, the renowned psychologist, discussed the nonrational aspect of the human being and people's ability to examine their inner selves.

These are just examples to show that EI has played a role in people's lives from ages ago, but only now is it being discussed in a structured way. In fact, organizational theorists joined their psychologist contemporaries in the twentieth century to try and assist people in understanding and solving interpersonal issues as well as their own internal issues. However, it was not until the 1980s and 1990s that experts began to take EI seriously as a unique intelligence component.

Researchers such as Jack Mayer, Reuven Bar-On and Peter Salovey, whose works on EI are among the most prominent, built on the findings of people like David Wechsler and R. W. Leeper, who began by identifying "affective intelligence" in the 1940s, and acknowledged it was a significant part of general intelligence. The two were confident in their assertion that affective intelligence, which has to do with emotions raising political alertness, is quite distinct from other forms of intelligence, namely intellectual and cognitive intelligence.

Why the Concept of EI has Gained Currency

Goleman pointed out the changing work environment, where traditional rules of management are no longer fruitful. He says people in general are being judged differently, and that the number of academic or professional certificates they hold, the kind of training they have or how smart they are considered to be is only part of it. On top of all those criteria, individuals need to demonstrate that they can handle themselves well and relate well with others. That last bit involves EI, and that shows how much value organizations are attaching to relationships at work.

Why the shift, one might wonder? The reason CEOs are taking EI seriously is that, with continued globalization of almost everything, competition is becoming stiffer and the growth of advanced technology is not helping matters. At the same time, pressure is increasing for people to balance work and family, and in the process the performance of several companies seems to be suffering. In trying to improve systems, management has realized that having highly qualified staff is not enough to run a globally competitive organization. Relationships within systems need to be nurtured, and it can only be done with the help of people with high EI.

At the same time, it has been discovered that people's IQ can be adversely affected if they are overwhelmed or stressed, and therein lies another reason for managing personal emotions. In fact, relevant studies have shown that stress is detrimental to a person's thought process, which means the affected person cannot be relied upon to make helpful decisions. Modern CEOs are wary of having employees who are stressed because they can easily damage not only their own health, but also their relationships with coworkers and other people they relate with in the course of duty. As such, such employees will have a bad impact on the company's bottom-line.

Experiment on the Impact of Emotional Intelligence
To explore what the accompanying benefits would be, Motorola decided to teach its employees fresh ways of handling tension and withstanding conflict. The company contracted by Motorola to conduct this experiment was HeartMath, LLC, a company in Boulder Creek, California, offering consultancy services.

The core of the six-month study was to establish whether the company would benefit from using management techniques that focused on managing emotions to a level where productivity would increase, teamwork was enhanced and communication and health improved. The highly awaited results were encouraging, because 26 percent of the company employees under the training program had their blood pressure drop in a healthy way; 36 percent had their stress symptoms reduced; and 32 percent said they felt more content.

What was even more impressive from a business perspective was that 57 percent of the participants' level of productivity rose by over 50 percent. Forty-seven percent said teamwork had been enhanced by 50 percent. In the meantime, measurable quality of output increased by 22 percent just from the training. This was clear evidence that emotions play a crucial role in the performance of organizations, and it makes great business sense to leverage them.

It has been observed that smart people sometimes make mistakes you would not expect from them, mistakes that warrant being labeled stupid. An article featured in one of *Fortune's* 1999 publications indicated that many CEOs who end up failing have encountered management difficulties mainly due to lack of emotional strength. In short, being smart or having a vision is not sufficient to bring success. Emotional intelligence and strength, which go hand in hand, are necessary.

Signs of Emotional Intelligence	
Related to Self	Related to Relationships
Being self-aware	Being empathetic
People with high EI can read their internal state, know their abilities and acknowledge their preferences. This means they:	People with high EI are aware of other people's feelings, concerns and needs. This means they:
have emotional awareness	understand other people
	are capable and willing to

can make accurate assessments of their strengths know their limitations have self-confidence.	develop other people can anticipate other people's needs and are ready to help them with those needs are capable of leveraging diversity are able to read emotional currents within a group and easily understand power relationships.
Being able to self-regulate This means the person is able to take charge of whatever is going on internally, including impulses. These people are notably: trustworthy capable of self-control, so that they can tame disruptive impulses conscientious easily adaptable to change receptive of new ideas and techniques.	(2) Having great social skills These people are great at getting the best in other people. They are able to: influence others positively communicate clearly and effectively manage conflict lead well catalyze positive change build bonds between people collaborate and cooperate well with others

	create and enhance group synergy.
Having motivation This means EI people tend to move towards their goals. This is because they: have self-drive are committed to success, be it for a group or an organization take initiative are optimistic even when there are challenges or setbacks.	

Liaison Between Head and Heart for Success

For an organization to succeed, the people making crucial decisions or leading significant sections need to have high EI on top of a high IQ. Although many established organizations may not have known before about the implications of having people of high EI on their team, they need not worry about overhauling the management team. This is because intelligent people can be trained to acquire the levels of EI necessary to take the organization to great heights. Staff can be trained, but they also need to practice what they learn on a daily basis so that it becomes entrenched in them as if it is part of their natural talent.

Why it is Possible to Learn and Acquire EI

When corporate organizations hire people who have great EI, mostly they want someone who is "street smart" in terms of being able to use their IQ and their heart to make decisions that suit situations as they arise.

Three Main Components of EI

1) Awareness of one's own emotional condition

2) Awareness of other people's emotional conditions

3) Capability to manage one's self-awareness and that of others.

These three aspects of emotional intelligence are the ones that basically bring about positive changes in an organization, helping it leave the realm of high-performing companies to join the cream of top global performers.

Three Main Areas Impacted by EI

Owing to the self-awareness high EI employees and leaders may have, and their awareness about others, coupled with their ability to make good use of that awareness, organizations are positively impacted in the areas of:

(i) Personal development

(ii) Collaborative learning

(iii) Systems thinking

Basically, when it comes to personal development, trainers want to focus on increasing their trainees' capacity to gain insight into their various motivations and thought processes, as well as personal beliefs and choices. They also want to guide the trainees in learning from their own experiences. One manager, Debbie Reichenbach, has been quoted as saying that people are better at making decisions when they feel balanced.

That is the reason companies that are geared towards great success invest in personal development. They want to tap the potential of intelligent people by enhancing their emotional intelligence so that the head and the heart can work in tandem with each other to make decisions that are best for the success of the organization.

When it comes to collaborative learning, organizations looking to breaking success ceilings try to enhance harmony in the workplace so that teamwork can yield great results. Of course, in an organization like a company there are small systems within the big system, and it is important for each department to work cohesively if the entire company is to succeed and reap the most profits possible.

Still, there is another side to enhancing collaboration among staff. When people understand one another well and empathize with one another, conflicts become rare and the organization does not have to waste time and resources firefighting.

As for systems thinking, which is also impacted positively by EI, it benefits greatly when people are personally effective as well as when they collectively work well. It has been observed that visible patterns of behavior in an organization are a result of systemic structures that include beliefs, feelings and various forms of motivations.

Example:

A mother who misses her son's sports competition, where he is a major participant, can apologize for it. But suppose the son points out that this event is the fifth one his mother has missed this year. Would missing the sports competition be considered an event anymore? It definitely has been shown to be a pattern.

These are the kind of patterns organizations try to break by delving deep into the reasons for these behaviors. In the case of mother and son, a close family member who has high EI is likely to unravel the underlying reasons for the mother's unpleasant pattern of missing her son's school events. One reason could be that the mother feels the pressure of beating work deadlines for fear that, if she takes some time off from work, for whatever reason, she might miss deadlines, and in turn her colleagues might view her as being incompetent while her bosses might consider her uncommitted.

How to Practically Enhance Emotional Intelligence

Organizations take a long time to adopt most new concepts even when, in principle, they have acknowledged them as worthwhile. This is not very different from the concept of enhancing EI, especially in the context of systems thinking. The few organizations that have tried it have succeeded, and some of them have created training models to be used by their respective organizations. Others appreciate the importance of nurturing relationships among systems and are prepared to allow other organizations to use their training model.

In fact, the issue of implementing the great concepts of EI is alive within the corporate world today, and managers are continually identifying business areas that relate to training. Many of the areas in the corporate world that could do well with EI training programs include executive coaching, staff recruitment and retention and teambuilding as well as leadership development. Luckily, it is not always necessary to create a fresh staff training program from scratch. Like some companies have been doing, one can incorporate EI training into an already existing program.

Real Examples of Corporate EI Programs
(1) American Express Financial Advisors (AEFA)

In 1992, this was the first company to design an EI program that proved to be successful. The program was called the Emotional Competence Program. This program was carried out by a team led by the company director of leadership development, Kate Cannon. The underlying philosophy for the program was that if individuals managed to identify their personal motivations and acknowledged their emotional states, it would become easier for them to communicate with their colleagues in a more effective way and they could build stronger relationships. In the process they would achieve their personal goals as well as their professional ones.

According to Kate Cannon, the intention in designing the program was to assist leaders to propel the company to great heights in the modern world. She emphasizes the fact that a lot has changed in the world and companies cannot grow while still relying on traditional ways of running company affairs. She points out the need for leaders to be flexible and adaptable, and also to be prepared to develop relationships.

The Emotional Competence Program was termed a success because attendants were assessed and the findings were that they had gained a lot of practical knowledge from the program. Sixty financial planners who were among the trainees performed much better than their colleagues and they credited the program for their success. Managers who attended the pilot study also had great performance improvements. They had their sales revenues exceed those of their counterparts by a whole 11 percent. Not surprisingly, the program has continued to help professionals increase their companies' revenues.

The research that took place at AEFA was very encouraging, because it showed that people's emotional quotient inventory (EQi) can increase with appropriate training. This measure actually assesses a person's ability to manage everyday pressures and demands leading to great success. For the participants, who were part of the AEFA group, one test was conducted before the training and another one afterwards, and when the two sets of tests were compared it was evident the scores for the participants had improved on most of the 15 subscales on the EQi tool. Thus, it is advisable for companies that are serious about enhancing their human resources to take their staff through a similar emotional competence program.

(2) Nichols Aluminum

One of the companies that adopted the Emotional Competence Program designed by AEFA was Quanex Corporation in its Nichols Aluminum Division. Nichols Aluminum is an aluminum sheet manufacturer that has plants in Colorado, Iowa, Alabama and Illinois, and it has 750 employees.

One important thing CEOs can learn from Nichols Aluminum is that EI programs are not necessarily for failing companies. The program is great at helping organizations already performing well to go a notch higher, and even better, to sustain their exemplary performance. In 1999, for example, Nichols Aluminum's sales revenues had increased by 17 percent and reached $312 million. Yet the company appreciated the monumental task ahead in maintaining the impressive sales record. The company president at the time was Terry Schroeder, who had been with the company from 1996. He expressed his wish to empower company employees so that they could make even better decisions. The company's training manager, Jan Daker, had read about AEFA's program and recommended it.

For Nichols Aluminum, the aim was to enhance trust and improve communication among senior managers as many of them were laidback. In fact, the culture within the company was somewhat withdrawn.

One participant in Nichols Aluminum's 1999 program disclosed that they never knew how to go about expressing their personal emotions constructively, and so the emotional competence training came in handy in helping enhance relationships within the management team.

To assess how successful the program would be, the 20 people who took part in the program were given an assessment test before undertaking the program, and they were given another when the training program ended. The consultant in charge provided feedback to everyone who participated, which was designed to show the individual what their strengths were as well as to highlight the areas that required development.

After that, the consultant did quarterly follow ups with participants at Nichols Aluminum, not only to see how well the participants were doing, but also to keep them motivated. The participants acknowledged that they needed that emotional competence in order to continue to build their business.

The years following the training of Nichols Aluminum's top management have shown evidence of improvement, with the company's bottom-line being impacted positively. This has mainly resulted from improved relationships among fellow staff members. One employee, in giving feedback, said one colleague was now in better control of his emotions and did not manifest anger and frustration as much as before. That colleague was impressively reported to have become more people oriented, good at paying attention to other people and even asking for input from them.

The consultant also learned that, after the training, the senior management team appeared to be more united. The managers had led the different sections of the company in improving communication, which made the working environment much more pleasant. What was even more impressive was that in the company's Chicago plant, where just four people had undergone the emotional competence training, the impact the program had on them trickled down to the rest of the plant, and the positive results could be discerned from the rest of the 130 employees.

The company's Chicago site improved its financial performance as sales volumes increased and operations became more efficient. Due to the success of the first training program, the company training manager, Daker, was so delighted that he organized for a second training program in the summer of 2000.

These examples of well-known companies are meant to show that EI is not an academic concept or one that is still nascent. It has been tested, implemented and found fruitful. In fact, the success that comes with training staff through EI programs is immense mainly because, as individuals become individually more productive, and even more productive in liaison with others, they also avert losses. There are less resources spent on solving problems, because many would-be problems are nipped in the bud before they can take shape.

Best Way to Practice EI

The best way to practice EI is within a team setup, because not only will others observe and tell you when your behavior is changing for the better, but everyone will be able to notice when the bond among team members has strengthened. Company management can also notice when teamwork in a given department or section is bearing fruit, especially in reducing lost working hours and other resources while increasing output under a less stressful environment. However, as charity is said to begin at home, you can always begin with your own habits, even as you acknowledge you do not exist in isolation.

1) Design clear guidelines to be followed by everyone, associated with respectful listening and talking at meetings. For example, you may have a guideline indicating no interruptions.

2) Make a habit of trying to learn the motivations behind other people's actions or thoughts when it comes to ideas that are of interest to everyone in the group. For example, you could ask, why do you consider this method the best? Or, how do you feel about taking a break on this project?

3) Practice being frank about your personal feelings, such as fears, shortcomings, misgivings and so on. For example, you could say: this may be surprising to you, but I have not understood any of what you have just explained.

4) Be accepting of other people's opinions and experiences as well as needs, taking them to be important because the other people value them. For example, if the finance manager communicates his intention to cut your department's budget, and he explains the pressure he is under to cut costs, you need to understand his position

and understand that he is taking the proposed action because it is necessary.

5) Give other people room to explain their behavior, as opposed to sticking with only one possibility, which you may even find unjustifiable. For example, your manager might report late from his lunchbreak even when he said he wanted to meet you at 2:00 p.m. sharp for a serious discussion. Instead of deciding that he thinks your time is not important because you are his junior, consider that he might be caught up in a serious traffic jam; he may have encountered an accident along the way back to the office; he could have received an urgent call that diverted him to another place; he might have lost his phone and had no way of reaching you to reschedule the meeting and so on.

Caveat

Emotional competence training is not a panacea for all problems ailing companies and other systems. It is only one of the fruitful solutions that organizations need to incorporate into their processes. However, the fact that it helps in self-improvement makes it very handy, because in many other cases, people become defensive when their weaknesses are pointed out. Where EI is at work, the individual is personally critical of personal actions, and personally responsible for taking remedial action. Moreover, people not only try to improve themselves for their own performance, but also to enhance their relationships with other people. That creates an environment that is conducive for other good ideas to take root and thrive in a sustainable way. In short, although it takes deliberate effort to enhance EI, it is more rewarding than many other programs floated in the business world.

Group Practice

- In a group setup, determine how enhancing EI within the workplace can positively affect company profits through favorable leadership, staff retention, staff health and career development. After this, embark on developing EI skills by using the guidelines already provided in the book.

• Go through the EI competencies already highlighted, still within a group setup, and determine which ones are relevant to your business so that you want to accord them priority. Take this opportunity to also rank your department or section according to the EI competencies you have collectively manifested. In case you have identified some organizational weaknesses within the department, section or organization as a whole, point them out and explain how they are limiting the employees' EI capacity.

• Write down what your roles and responsibilities are in the organization, and then try and learn how you can improve your EI as an individual or with other colleagues. Identify the specific areas you would like to improve by enhancing your EI and then draw up a realistic action plan.

• Identify your best bosses as well as your worst, and then pinpoint the positive and negative qualities that have had the greatest impact on you. Classify those positive and negative qualities as either emotional or intellectual.

Chapter 8: Organizational Structures and Systems Thinking

Oftentimes, organizations do not get the results they anticipate. This then begs the question, are organizations really operated and designed to generate the results they get, as has always been the belief? Do you find yourself perplexed by the conflicting results? In this book we shall explore a number of concepts on why organizational design dictates the outcomes of an organization, and also look into alternatives, all this with the aim of creating an understanding of organizational structure.

To fully understand organizational structures and their impact on outcomes, you need to ask yourself a few questions: Is organizational design restricted to the conventional thoughts about hierarchy, or does it entail more than that? Do an organization's procedures, policies, management assumptions, beliefs and actions influence its activities? In the past few years, organizations and organizational structures have undergone, and continue to undergo, constant changes as managers seek to achieve efficient yet flexible systems. Increasingly, managers are revaluating and questioning traditional means of organization.

Organizations are traditionally designed in a hierarchical manner since it is believed that such a system is easier to manage. An important question to ask, however, is what is more important—the management or the organization's output? Would management still be of value if the hierarchical structure failed to make the organization easier to manage and was not results-oriented? If you designed a self-managing organization, would there be any use for management?

Let us consider a ship, for instance. What do you think is the most important function aboard the ship? Could it be the captain, whose job is to run the ship? Is it the engineer, who ensures that the ship is powered? Or perhaps the navigator, who plots the course? The answer is none of these. They are all equally important when it comes to ensuring the voyage is smooth, but the person tasked with designing the ship is the most important of all. Ongoing operations in an organization are not as crucial as the design on whose basis the operations are carried out, which happens to be management's greatest responsibility.

Organizational Functions Within a System

There are two distinct types of functions in an organization, both of which have similar intentions but are different in how they are carried out. These are:

1. Processes that start and end with external customer relations

Results from this function can be easily measured.

2. Process enablers

These functions do not start or end with external customer relations, but instead support the first process mentioned above. The results drawn from these functions are more subjective than objective, and they are more difficult to measure.

Matrix Organization

Within an organization, there are those processes that function independently while others rely on other processes to function. In each of these cases, there are pros and cons when it comes to the actual operations. The question you should ask yourself, however, is, do the advantages outweigh the disadvantages when the processes are combined or when they work in tandem with one another? Or are "stand-alone" processes more efficient? Another intriguing question is whether customer service is a support service or a process on its own since it begins and ends with an external customer interaction, and it is also a process that should be maximized within an organization.

There are independent, as well as combined processes, grouped to form procedures that fully involve external customer relations and others that do not is called a matrix organization, and it has a hierarchical reporting structure. It has been noted over time by organizational theorists that matrix organizations do not bring about desired results. However, the reason for this failure is the lack of full implementation of required processes, which makes these types of organizations destined for failure from the very beginning.

Organizational Restructuring Within Systems Thinking

Most organizations, if not all, have a hierarchal reporting structure, and most of their activities look like those in the diagram above. However, this is not the ideal activity flow for maximum efficiency in an organization. To bring about a better flow, the organizational structure needs to be reconsidered. A systems rule to go by is one stating that we should not fight the system but instead should change the rules and the system will change itself. Buckminster Fuller, an American systems theorist, said that instead of trying to train people on the right things to do, an organization should be designed in a way that doing what is right is just the path that promises the least resistance.

For you to lay down an effective organizational structure, you need to learn about those that are already in place and how they work, as well as their pros and cons. Let us look at the matrix system, beginning with a process. A process is made up of a number of functions, and the purpose of each of those functions is to optimize output. This means that the process is tasked with generating optimum customer satisfaction within the shortest time possible, using the least resources possible. The functions are then assigned to supervisors who report to a higher authority. From here on out, the structure begins to take a new form. Supervisors are responsible for making sure that all tasks are well completed.

Loopholes in the Matrix Organizational System

So far, from the structure above, you can identify a few problematic areas. The main ones are listed here.

1. Focus on optimizing individual components of the organization.

You cannot optimize the system by optimizing its components. Instead, it is better to enhance the interaction among components since that yields better results for the organization than when functions are individually optimized.

2. Supervisors can easily lose sight of what is important.

In this form of organizational structure, supervisors are likely to focus on supervising tasks and become fixated with pleasing the manager in a bid to improve future performance appraisals.

3. Managers report to a higher authority.

Just as employees are eager to please their supervisors instead of focusing on improving efficiency, managers who, in turn, report to a higher authority, also do the same. They strive to please their superiors, and that draws their attention away from the tasks they are supposed to have been performing for the good of the organization.

4. The structure is complicated.

The process requires support from several different functions for it to produce desirable results, and that complicates the structure even further.

Total Quality Management (TQM)

The next trial was using the total quality management (TQM) whereby every group focused on its customers' requirements. The single most common reason why the TQM failed was because it did not alter organizational structure. Both loyalties and reporting structures remained the same, diluting any efforts made by TQM. Additionally, TQM placed its focus on the immediate customer rather than the whole organization.

How Self-Managing Workgroups Operate

In the recent past, there has been thorough research and application of self-managing workgroups or semi-autonomous teams. This approach has been successful in many organizations but has also reported failures. Most of the failure experienced has been attributed to lack of change in the overall system structure. Another cause of failure is the confusing messages sent out to these groups, which frustrate their potential success. These mixed messages mainly come from management, whereby they award individual performance, on one hand, as they praise teamwork on the other, leaving workers confused.

For the success of a multi-function process, every function must concentrate on both its operation and overall results. Functions should ideally operate under a self-managing workgroup, evaluated according to its contribution towards the overall process. This setup brings balance between function and process whereby a function fails if the process fails. Constant communication between each function and the whole process is vital for the progress of all functions.

In self-managing workgroups, there is a difference between value and importance. Perhaps you wonder why it is that some people are paid more than others in the same workgroup and yet each member is important. The answer lies in the value of each skill. The value of an individual is based on how common or rare the skill is, how much time it takes to acquire the skill and the level of expertise in that given skill. An individual who is easily replaceable is not highly valued. A good example is that of a surgeon and his assistant. Using the value parameters mentioned above, while both positions are important, a surgeon's job is more valued than that of his assistant.

How to Eliminate Bureaucracy and Organizational Politics

Bureaucracy and politics in an organization are greatly promoted by the traditional system structure. As the first principle of systems states: structure influences behavior. Hence, it is an emergent quality of the structure. Few, if any, people know how to tackle the bureaucracy that plagues systems. It is also human nature for people to do what they feel makes sense at a given time. With this knowledge, we can alter the structure in such a way that bureaucracy brings no benefits, thereby discouraging people from taking that route.

Reengineering Organizations

Reengineering seems to be by far the most promising approach to "curing" organizational problems. One of the best works that give comprehensive insight into this is Dr. Lawrence Miller's *Whole Systems Architecture*. In his work, he points out that for people to bring about long-lasting change, they must change the entire organization structure and not just the reporting structure. All components of the structure, including policies, processes, rewards, incentives and management philosophy need to be changed, rather than just blocks on the chart.

However, the term reengineering is not suitable for most organizations since many were formed out of necessity. You, therefore, cannot reengineer that which was not initially engineered. The organization's structures and functions were formed naturally on a needs basis, and all other developments within the organization have been modified to fit into the structure. This is how most organizations evolved and developed over time.

Changing the Reporting Structure

Naturally, it can be tough to transition from one way of doing things to an entirely different way. An organization should offer support to self-managed teams as they transition from the orthodox supervisor-employee relation. For this reason, let us create a position called senior associate. This position is not meant to replace that of a supervisor but instead to provide development support services to the team. The senior associate provides consulting services to the function and is also evaluated by the same function. This ensures that he is focused on the function assigned rather than serving his manager, since he is not subject to one.

Another individual who can be put in place is a process coordinator. He will be responsible for providing services to the process functions and the senior associates. As such, he will also be responsible for the results of the process and development of senior associates as well as performance of process functions. The process coordinator reports to everyone in the process. Note that he does not take up traditional management functions such as organizing, planning, controlling and directing. He is evaluated based on teams' and senior associates' perceptions and the results of the process.

You are probably asking yourself, how should we deal with the challenges in supporting services and management under a traditional structure? The answer lies in the reporting structure. Just like the senior associates and process coordinators mentioned above, who report not to higher authorities but to those they supervise and serve, support services should be evaluated and also report to the very functions they support. This way, supportive systems are kept in check, and they focus on functions.

This solution, however, has its downside. Support offered at no cost to functions for their enhancement and overall process results can create exploitation of these support services, leading to an imbalance. To cushion this, individual functions, as well as their processes, must be charged a fee for using support services. This moderates the use of support and dilutes the results of the process, consequently making the process self-limiting rather than management-limited.

Similar to process coordinators and senior associates who support and report to functions and their processes, there are also senior associates and service coordinators in service operations. Process coordinators, service coordinators and upper management team up to continually develop the design of the organization which is now a self-directed, self-managing, self-evaluating and results-oriented system. The main role of the senior associates is to focus on long-term time frames since self-directed functions focus on their tasks in short time frames. Coordinators are tasked with coming up with an even broader perspective on overall processes, and to consider longer time frames.

At this point, you are probably asking: How shall we transform from the present traditional system to the envisioned operation? Is it easier to discuss it in theory than it is to implement it? The secret lies in gradual change, rather than attempting to change the whole system at once.

Many organizations try to change the system all at once, such as in many TQM implementations, but end up failing terribly. The traditional approach has always begun by convincing top management and then proceeding to involve and mobilize the whole organization toward change. Organizations eventually become numb to change interventions, owing to the numerous attempts that failed in the past.

As pointed out before, structure influences behavior. The best approach to organizational change is the one-rule-at-a-time approach. Procedures, policies, incentives, rewards and the rest all make up the structure, and consequently, the rules of operation. For you to change the organization's operations, you need to change the rules. You need to do it one rule at a time, and in response, the organization will change itself.

The basis of all initiatives, which include introduced concepts and transformational plans, is measuring what is needed. This step involves seeking the voice of three most important players in an organization: the customer, employee and business. Through interaction with the three voices of the system, you need to first gain an understanding of what is needed, and then proceed to develop systems that support those measures. Processes will tend to be objective while support services are more likely to be subjective.

Empowering the People Through Groups

A smart approach to transforming different segments in an organization is giving power to the people by working with groups within the organization. Pick an organization, introduce a set of ideas to groups, create an understanding of the concepts, involve the group and also work with it to coin its very own transformation plan. In so doing, people will commit more to the transformation plans since they are part of them, as opposed to feeling obligated to enroll in someone else's plan. It is, however, more appropriate to begin working with an organization's processes before moving on to support service groups.

Change is not a one-time objective to be accomplished, but rather a never-ending journey. With this in mind, implementers and receivers of change should not be in a hurry to move from one point to the next. When implementing the change in groups, it is best to deal with one group at a time. Once one group becomes capable of functioning by itself, you then proceed to the next one and so on. There should, however, be enough support for the groups throughout the entire transformation process. The assistance should be aimed at nurturing, supporting and coaching rather than gearing up to reach a destination. Remember, like we said, change is not an objective but a journey.

There are questions that a group needs to ask itself at the beginning, which can assist in shaping its transformational plans. The questions will not only guide the process but also enable the group to evaluate itself and, at the same time, provoke great ideas. These questions include:

Questions to Help in Group Transformation

1. What commendable things is the group doing that it should continue to do?

2. What areas need improvement?

3. What should the group start doing that it is not already doing?

4. What should the group stop doing?

The process does not end here. There is still a lot to be done. Remember, this is only an idea that has been laid down and has not been actualized as yet. The descriptions above do not make the process happen but rather enable it to happen. You must be careful not to have your ideas become a fad, simply applying formulas without much thought put into the process. In an organizational context, there are no constant formulas or solutions to anything as there are too many variables involved, and so no single formula can work. Constant assessment of a situation in an organization is the safest way to proceed.

Let us go back to the initial question on what the next steps are after developing the ideas. Here we shall look at a set of guidelines that can assist in drawing specific approaches for different phases of implementation.

Step 1: Alignment

This strategy has been used in a number of places. One can only align to something if there is a focal point or something to align upon. For example, a laser has the capacity to align because it is focused through a lens. In the case of an organization, a set of statement of beliefs can be written down on a piece of paper by the head of the organization with the aim of creating a focal point for alignment. The statements, which are best written under a persuasive title, should represent the voice of the customer, employee and business. The paper should then be shared all around the organization for awareness. Leadership should set an example for the rest of the organization by acting in line with the written beliefs.

Step 2: Create a role model

The next crucial step is working with one group, transforming it to the desired standards and using it as an example to the rest of the organization. This aims to show the rest of the groups that changes are possible. Transformation of the group should begin with an individual, then the group and later the entire organization. Developing a sense of responsibility and ownership within the group drives change through the whole organization.

Step 3: Set up meetings

You should then schedule meetings with each individual. The aim here is to create a sense of value in every person, giving them confidence that their input is important. The set of questions here can help set the basis for discussion:

1. What commendable things should the team continue doing?

2. What areas in the group need improvement?

3. What should the team stop doing?

4. What things should the team be encouraged to start doing?

5. What should the team expect from me?

6. What should I expect from the team?

The most important agenda in this meeting is to gather information regarding each person's expectations, and to help you create your own expectation of them. At no point should you argue or debate the individual's views. You can also coin questions of your own to draw clearer answers from questions from the set that were not thoroughly answered. It is also okay to ask for elaboration where necessary. Remember to record notes on each question asked.

Step 4: Form small groups

After the one-on-one with each individual, you should have people meet in small groups of not more than six to discuss the answers derived from earlier meetings. The aim of these meetings is to have people discuss and agree upon the most important questions, which are vital in shaping the organization. This creates unity among the individuals, which happens as they put aside their personal differences and share insights. At the end of the meeting, the small groups subconsciously produce a list of vision, purpose and values, which are of greater quality than any set you can gather from asking questions.

Step 5: Presenting the Findings

It is time to present the findings of the small groups to the larger groups, and here everyone has a chance to weigh in on the findings. The entire group then comes to an agreement on vital matters and decides on what the next steps will be with regard to the results. Some of the tasks agreed upon are taken up by the group, while some are outsourced to other sectors of the organization. The other sectors are responsible for supporting the tasks allocated to them, and they should do so without fail for the full realization of the group's end goal as well as to avoid discouraging the group.

At this point, the group can make its own decisions on what tasks can be accomplished and when. Individuals are also allowed to take up leadership roles and become responsible for progress in certain areas. This now becomes a team-based self-managed group, which is the aim of the whole lengthy process.

It is now your responsibility to support the group in whatever ways are possible by offering yourself as a resource. Note that you will not do the team members' job for them, but instead you will act as a coach, coordinator and facilitator, since all other responsibilities under the traditional structure such as organization, planning, controlling and directing have been assumed by the group. In such a situation, the position of senior associate can be created during the process as there is currently no established model for senior associates to emulate. It is therefore vital that this role be developed.

Chapter 9: Introducing Systems Thinking in An Entity, the Simple Way

Most leaders are afraid of learning new approaches to problems, especially when they think the models they are using are working. Introducing the concept of systems thinking may require you to wait for the opportune moment, otherwise it is likely you may meet stiff resistance. One of the best times to introduce the systems-thinking approach is when leaders are facing a certain challenge and all their approaches have failed. At such times, leaders are often ready to learn about other possible ways of approaching an issue.

Such are the opportunities you should take to introduce systems thinking, because at such times an organization's leaders are not only psychologically prepared to listen to new ideas, they are also ready to bring everyone else in the organization on board. For the many years that systems thinking has been taught and practiced, the following methods have been successfully used in teaching the concepts:

(1) Drop-in or ad-hoc method

This is applicable where one introduces systems-thinking concepts after all other approaches by the business leaders fail. The method can also be used where a particular problem has been recurring and is almost perennial. In such situations, members are often already frustrated by previous approaches and they are anxious to learn about the new method. When using this method:

- Use a simple loop diagram as opposed to one that is complex.

This way, not only will you be able to communicate your message more easily, but you do not risk scaring off your audience with complex diagrams.

- Use simple language that is easy to understand.

Using simple language is advisable for the same reasons you should use simple diagrams. In the same vein, initially avoid systems-thinking jargon. This is because using complex diagrams and jargons from the outset can put off the audience, who may perceive the subject as too complex, and this will hinder effective communication.

(2) Use of tutorials

In this method, basic elements of systems thinking are introduced after having been customized to fit a specific business model.

(3) Conduct workshops

For bigger organizations, you can liaise with them to organize workshops on systems thinking, where the learning objectives are:

(i) To have an understanding of the concepts and tools of systems thinking.

(ii) To learn how to employ systems thinking in solving various problems and uncovering leverage points. The leverage point is a point in a model where a minor change may have a big impact on the entire system.

(iii) To increase your ability to apply systems thinking in solving different issues.

What a Systems-Thinking Workshop Involves

• You need to start by introducing the concepts and tools of systems thinking. This will help in giving the participants a systems-thinking approach to various problems and will help them to uncover the leverage points.

• Take a single problem and discuss it as a test, so you can see whether systems thinking would work well in that organization as it is. Probably, more consultations with other stakeholders might be required.

- Come up with suitable solutions to use in solving the identified problem.

- Assess what situation to solve next depending on the outcome of the workshop.

Practical Example

This method of using workshops was used by a cable company that installs internet to their clients. Initially, the company used technicians to install their customers' software. However, in a bid to cut down on the number of technicians the company needed for those installations, the company decided to give incentives to those customers who were willing to install the software on their own. This meant the company would save expenses on account of reduced number of technicians.

This was successful as many customers, especially those who were tech-savvy, decided to do the software installations on their own. This freed up almost half of the technicians who had initially been involved in software installations, and although they were not laid off, they were able to render their services elsewhere in the company. In fact, they were redeployed to serve in other departments where their contribution was expected to raise productivity.

Gradually, the number of self-installers increased. However, most of the self-installers were less tech-savvy, compared to those involved in the initial installations, and they required assistance. This led to the need for more technicians to go out and assist the customers, hence returning the cycle to where it was in the very beginning.

When this cycle was analyzed by the group, which had been divided into several teams, each looking for the loop, people realized this was a balancing act being caused by one delay. The team was also able to uncover the leverage point, and they decided to come up with better strategies to ensure that their customers were able to successfully install the software all on their own. As part of the incentive for self-installation, a package for technical education was also included.

- Use of models

Models are useful in systems thinking because they represent a plan that is easy to understand and one that can be applied in various human situations. After understanding a model, one can easily identify a problem that can be solved by use of that particular model.

- Organizational assessment

This includes checking what processes are working and which ones are not within an organization. This should then be represented in the form of a diagram for easier understanding. The problem is then identified by participants and this makes it easier for them to come up with outstanding solutions.

This method was successfully used in an organization where the board of directors and the executive director had a broken relationship. A loop diagram was created to show what an outsider's opinion of their relationship was, and what the effects of that relationship were. Note that when such an analysis is done, it should be expressed in a manner that does not assign blame.

In this instance, the analysis led to the beginning of an open and truthful discussion, which eventually created more trust, openness and role clarity between the executive director and the board members. It was revealed that the quality of relationship between the board and the executive director had declined over time, and that had led to loss of trust and declining openness. This had then led to the board members being less interested in contributing and raising funds to run the organization.

All these variables, when represented in a loop diagram, show the impact each element has on another variable. The approach helps everybody to own the problem rather than shifting blame to others. When introducing the system loop diagram, it is important to clarify that it represents only one opinion, which incidentally might be right or wrong, but most importantly, that gives a starting point for the discussion.

Summarized Guidelines for Teaching Systems Thinking

a. Don't use too much jargon.

If you use too much jargon, it can quickly put off people who do not understand it. Use simple, clear and familiar language that people can relate to.

b. Use real-life examples to explain various concepts.

Relatable examples are helpful in communicating how systems thinking works. The example of Benihana's restaurant service, included later in this book, is a good one, because it is simple and people can connect to the situation described.

c. Appreciate other people's knowledge and experiences.

It is not advisable to dismiss everyone else's ideas just because you think those people are less knowledgeable or have inadequate experience. Appreciating them boosts their self-esteem and helps ready them to learn. There may be a good point behind what they are expressing, although it may not appear like it at the surface level. Lead them with questions to see if you can unravel some important information about the organization.

d. Look for a problem that has persisted in the organization.

If you can identify a problem that has been persistent or recurrent in the organization, people will be keen to take a look at it from a different perspective, and the problem will serve as a good demonstration that systems thinking does actually help solve organizational challenges.

e. Help people see the effects of a problem in a loop diagram.

This helps to determine the problem's starting point.

Chapter 10: Systems-Thinking Models

When thinking in systems, you need to make use of models that identify existing bottlenecks, and then find leverage and make use of feedback loops. Before that you need to have decided the goals you want to achieve. Something else you need to appreciate is that in order for great goals to be met, you need to have a good working system. If you already have a system in place, you need to study it keenly to see if it needs to be modified in order to correct its weaknesses while also strengthening its good points.

In short, you need to aim at changing your default behavior and that of the other people involved with sustainable changes, so that the system can continue on a successful path as long as your goals remain as you have set them. To help with this kind of success, you can use any of three mental models best used in systems thinking, which play a central role in designing an approach towards growth. Those models help in identifying bottlenecks and in creating leverage as well as in providing feedback loops.

In fact, when you make use of these models within systems thinking, it helps you analyze every part of your personal life, and also enables you to make appropriate improvements. Experts describe that capacity in systems thinking as a paradigm capable of shifting every other paradigm. This shift does not only occur in an individual's personal behavior but in every system where said models are used.

Some people might wonder: is it not enough that you already have goals you are motivated to work towards? While having motivating goals is great for success, such excitement can only keep you focused for a limited period; usually a short period.

Take the example of you and your friend whose only physical exercise is a walk in the neighborhood with your dogs. If your closest neighbors have decided to run a marathon and they encourage you to join them, you are, very likely, going to begin a running exercise regime to ensure you are fit by the time of the marathon. However, do you really see yourself continuing with that disciplined exercise regime after the marathon? Very unlikely. You ran regularly to prepare yourself for the marathon, but now that the marathon has come and gone you no longer have motivation to continue running.

The takeaway message here is that the life of systems and processes is always longer than that of motivation.

Consider that a system comprises other smaller systems, which are often taken to be subsystems, whereas in daily life people like to look at them as constituent parts of the big system or the whole. Although those constituent parts are unified and directed towards achieving a common goal when merged into one unit or one system, the properties of the whole system are different, much greater and more impactful than those of its constituent parts added together.

Also, when one of those subsystems is altered, either for better or worse, it affects the main system in a bigger way. In fact, it is not always possible to predict how the big system is going to be impacted by a tiny change in one of its constituent parts. Experts explain it as the system displaying emergence.

Everything Happens to be A System

You will understand the universe better when you consider it to be made up of numerous systems, where there are numerous interconnected systems that not only intersect but also interact with one another. People's actions have an impact on those different systems, sometimes consciously and other times unconsciously.

If you want to appreciate these connections and your role in the system, think of organizations you are directly part of, and communities you are associated with. Do you realize they affect other systems in their own ways? And since you are part of them, it means your actions have an indirect impact on those other systems. There are other systems that people rarely consider to be linked to them, but they are. These are systems like vehicles, public transport systems, the stock market and everything else that crosses your path either by physical presence or the impact of your actions, both minor and major.

Individuals as Systems

To help improve performance and create a healthy existence both as individuals and as social groups, it is important to understand that individuals are systems, too, with other constituent systems within them. One of a human being's important systems is the biological system, which in itself has numerous smaller systems, such as the digestive system, the muscular system, the skeleton system, the endocrine system, the nervous system and others. These, in turn, have smaller constituent systems, and that complexity of systems continues up to the person's cellular level, moving on to the molecular level and even to quantum levels.

What is interesting about human beings is that they have the ability to think at an abstract level, and so they are able to rise above the constituent systems of their being as opposed to being trapped within them. Human beings are also adaptive, and so they keep evolving. In the process, they are able to rid themselves of many problems before they become serious and inhibit the working of the human system.

In short, human beings, unlike other systems, do not wait to take instructions or to be coded in a certain way so as to modify their way of working. Rather, they are able to proactively tailor their path of operation to suit their needs.

It is advantageous that people can change goals they had previously aligned themselves with, and consequently alter the way their systems work. People have the capacity to select their preferred goals, and to generate different ways for various systems within them to operate so as to meet those newly chosen goals. People's abstract capacity enables them to even change their belief systems and other deeply held values. That separates people as systems from other systems.

Most Important Systems-Thinking Mental Models

1) Bottlenecks

This model, which is also referred to as the theory of constraints, has been adopted from the manufacturing set of systems. It acknowledges that every single system has its limitations, basically because there are constraints, but there is normally one distinct constraint that is tougher or more sensitive than the rest. To understand that better, think of the analogy of a chain and its weakest link. There is always that part of the system with the greatest weakness, which, if strengthened, will make a world of difference to the way the entire system works. That most impactful constraint is what is referred to as the bottleneck.

If you are considering the system as a process, the bottleneck is that part that has the greatest congestion, which is the main cause of delay in producing expected results. As Eliyahu Goldratt explains in his writings, a system's performance is mainly limited by the system's bottleneck's output. In other words, even if you make positive changes in various parts of the system, it will not operate optimally unless you make changes that impact the bottleneck.

One important thing to learn from the theory of constraints is that people and institutions waste a lot of resources addressing system areas that have no bearing on the bottlenecks. Even when individuals focus on self-improvement, they often address areas that do not change their overall performance or capability. From this systems-thinking model, you can see that it is not effort, per se, that is required in order to improve efficiency, but rather effective application of the effort.

It is important to note that the removal of a bottleneck is not a goal in itself but a way of solving a problem that is inhibiting the smooth running of a system. As such, the focus should remain on keeping the system running unhindered, otherwise there is always a chance of a bottleneck developing elsewhere if the system is not well monitored.

At the same time, you can only manage to identify other problems in the system with clarity after dealing with the bottleneck. As an individual, once you remove the bottleneck inhibiting your performance, you are most likely going to experience overall change in your life, including your level of happiness and motivation. Normally there is a phase transition in a system following the removal of a bottleneck, which impacts everything about the system. The best advice when you want to make meaningful changes in your life is to identify your bottleneck, leave everything else about yourself aside momentarily and focus on removing the bottleneck.

Practical Examples of Bottlenecks

Take the example of a burger house thata man named Baldwin visited after hearing about its delicious burgers. The outlet was average in size and had a clean kitchen that was visible to the customers. The burger ingredients were of high quality, but the range of choice was narrow. However, the place was cozy and the service great.

Months down the line, Baldwin visited the outlet again, and this time, too, he was happy to be there, going by his experience at the cashiers' section. When a customer stood at the cashier's line, the wait was short and the flow of people being served was continuous and smooth. Customers could even see from the line what the burger choices were, and so there was no time wasted once they reached the cashier's place. You would say what your choices were, either hamburger or cheeseburger, for example, and then state what you wanted added, such as lettuce, tomato, mustard, pickles and so on. If you wanted soda or juice you would state that as well, then you would make your payment and step aside for the next customer as you waited for your order to be prepared.

But, lo and behold! That was as far as the smooth sailing went. There was a disorderly crowd of people waiting for their orders to arrive, and from some people's restlessness, it was evident they had been waiting for quite a while. In the meantime, one could see some people working behind the counter as they frantically tried to assemble the customers' different orders. It was frustrating for some hungry customers to have to wait for so long, only to receive the wrong orders. So it was chaos in the kitchen area and frustration in the waiting area.

The outlet owners must have been excited about the rising customer traffic, but they must have been disturbed by the fact they could not render their services as efficiently as they had before. In fact, some people in the queue, waiting to pay and order, ended up leaving after noticing the growing crowd of customers. Such customers are unlikely to return to the place soon. In short, at the end of the day, with all the discomfort on both the customers and employees, the burger house's sales revenues cannot have increased significantly.

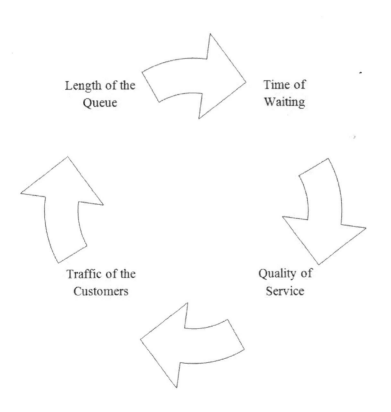

Baldwin lived in the neighborhood, so he had another chance to visit the burger house after another couple of weeks. As he joined the cashier's queue to order his burger, he noticed only one cashier was rendering services this time around, as opposed to two, the way it was the other times he had visited. One of the two cashier tills had a sign showing it was closed. On inquiry, Baldwin learned that one cashier had called in sick that day. What transpired was unanticipated but amazing.

Apart from customers waiting a little longer in the queue to pay, no chaos developed at the cashiers' section. Instead, Baldwin got to the cashier and paid for his burger. As he proceeded to the waiting area, he could see there were people behind the counter searing burgers and frying accompanying potatoes well ahead of time. In short, there were no frantic movements, as had been evident the last time. It was clear the kitchen team was keeping up with the flow of orders seamlessly.

In fact, at the waiting area only a couple of people were waiting to be served. That did not imply that sales had dropped. All it meant was that the system was more efficient. What had management done to improve matters? Well, they had not actually done anything when one cashier unexpectedly failed to report for duty. That then meant the flow of orders to the kitchen proceeded at a slightly slower pace. The efficiency and swiftness of the cashier section had earlier created a bottleneck within the kitchen, and that had made the burger house appear disorganized.

This time around, the people preparing the burgers assembled them in a synchronized way and at a good pace, which meant customers' orders were dealt with promptly as they arrived, as opposed to piling up as the customers crowded in the waiting area. In short, in a business or social environment you do not necessarily need to increase manpower, equipment or make any additional investments so as to streamline operations. You just need to identify the system bottleneck and clear it. In the burger house example, the bottleneck became clear in retrospect after the system accidentally streamlined itself when one cashier didn't come to work.

To be clear, in the case of the burger house, there was a bottleneck where a number of cooks assembled burgers and compiled the orders. This was because the burger ordering system that had two cashiers worked optimally, but that meant throwing too much work at the inventory system at once. Instead of being happy that the overall system was going to benefit from numerous orders, the kitchen staff was frustrated because the burger supply system was overstretched.

Where one system or subsystem develops a bottleneck, it is difficult for those involved to focus on the good of the entire system, owing to their misery. As you may imagine, there is also a great possibility for the output to be poor. In the case of the burger house, some customers who never ordered a pickle on the burger got one anyway and that resulted in waste when they didn't want the mistaken order they then received. Others who noticed mistakes in their order before digging into the burgers had them swapped, but that meant extended delays for other customers. Overall, therefore, the system with only one cashier that worked like clockwork could lead to sales of more burgers because it was a more efficient system, eliminating a troublesome bottleneck.

Problem of not Identifying and Acknowledging a Bottleneck

Let us use the example of the burger house again.

What transpired when two cashiers were accidentally reduced to one?

- The rate of receiving orders was reduced, because only one cashier was taking customers' orders.

- Consequently, there were fewer orders reaching the kitchen at a time.

- The members of the kitchen staff were able to clear the orders quickly.

- They were also able to assemble the burgers with ease and to pay keen attention to what the customers had specifically ordered. In short, the quality of the service was higher.

- No crowd formed in the waiting area because the quick pace of clearing orders was sustainable.

In short, the service at the burger house improved and many of the systems involved ran smoothly. All that great experience only cost the customer an extra couple of minutes to wait in line at the cashier's.

However, because the management of the burger house had no knowledge about bottlenecks and was stuck in the traditional method of running systems, they did not identify the solution the missing cashier had inadvertently brought about. So, when the cashier became well and reported back to work, the system went back to chaos, with tired kitchen staff nearing confusion and a crowd of customers getting frustrated with long waits and wrong orders.

Apparently, the only thing that might have drawn management's attention and prompted them to analyze the system to see what was wrong was if there was a suggestion that they invest more money, probably on manpower, or if there was a noticeable drop in sales. As for the causality of the events—the ordering and the serving—it was clearly counterintuitive.

Advantages of Identifying and Acknowledging a Bottleneck

Sometime in the travels of Baldwin, while walking around at an airport waiting to take a connecting flight to his next destination, he noticed a food outlet that was part of an international chain. There were around five cashiers at the counter of the food outlet, which also had a burger frying area, a burger assembly section, a potato frying area and a salad assembly section.

While the rest of the food was assembled in other sections, any soda ordered was provided by the cashier in addition to taking the customer's order. As Baldwin was about to be served, he heard a loud voice come from the back. It was the outlet manager yelling for one cashier to go to the back to lend a hand. So one cashier closed her position, washed her hands and joined the staff at the back who were frying burgers. Instead of three people frying burgers, there were now three, and the burgers began to flow from the burger station much faster.

Earlier on, the speed at which burgers were being supplied from the frying area had slowed, and paid customer orders had begun to accumulate. Luckily, the manager knew how to identify a bottleneck and how to clear it. When he identified the bottleneck as the burger frying system being overwhelmed, he decided to call for backup from the ordering system.

Some managers and employees might be of the opinion that someone should not change their work station when they are good at what they are doing, but when thinking in systems, resources, including manpower, need to be deployed where their services are most valuable.

- Why was the burger frying staff overwhelmed? They were overwhelmed because the cashiers were churning out orders at a very fast rate.

- Why did the manager not simply ask the cashiers to stop taking more orders for a while? If services at the counter had been halted altogether, many customers would, very likely, have left the queue and sought to have their meal at another food outlet. At the same time, the cashiers would have been paid for downtime, which is not very economical for business.

- Why was one cashier called to the back? By asking one cashier to lend a hand at the burger frying station, efficiency at the location improved, and that essentially removed the backlog of orders pending.

In short, every member of staff was being productive, customers both at the ordering section and at the food receiving area were kept hopeful and happy and the overall system worked impressively well. This was because the manager had not waited to be prompted by chaotic scenes of frustrated customers but had instead monitored the system as it worked and was proactive when he identified a bottleneck.

Valuable Employees Attribute Whatever Their Specialty

What could have happened if there was no cashier who knew how to fry burgers? Well, probably the cashier would have relieved someone else at the potato frying section who, in turn, would have joined the staff frying burgers. The point being made here is that it helps in business to have employees who can do more than one thing. One might not be efficient in everything, but at least it helps to have more than one skill related to the system within which one is working, and to have the right mindset.

In the case of the food outlet, 10 or 15 more minutes of piling orders might have caused chaos and probably a loss of customers. So, however one can help, it is important to build a culture where the staff appreciates teamwork and understands that the system can only succeed when every section succeeds. In traditional management systems, departments compete against one another and hardly appreciate that the overall system loss affects every subsystem just as negatively.

It is important that everyone involved knows that as long as the organization's activities involve producing a product or service from one end to another, such as delivering a customer's food order, the more of the activities an employee can perform, the better for the system's throughput. In short, whether an organization accomplishes this through in-service programs or other methods, it is always helpful for the system to have cross-functional individuals.

That is the reason knowledgeable people who can afford to engage house help also give them driving lessons and help them acquire driving licenses. When they are thus equipped, they can help pick up the children from school and drop them at different places for extracurricular activities instead of the parents having to take time off work to play chauffeurs.

When it comes to weaknesses within systems in general, it does not matter which department has the bottleneck. Unless it is addressed, the entire system suffers.

If you optimize a stage before a bottleneck, just because you can, for instance optimizing the ordering system while the burger frying system still has a bottleneck, orders will accumulate and that will result in more pressure being put on the staff working at the burger frying area. That will, most likely, lead to the staff serving poorly prepared burgers, and declining product quality is, obviously, bad for business. In short, optimizing what is already working while leaving the bottleneck unattended only worsens the case for the entire system.

Even optimizing a system or subsystem that comes after the bottleneck, without addressing the bottleneck, will still produce bad results. In the case of the food outlet, if the burger frying staff had remained overwhelmed but management chose to optimize, say, the burger wrapping or serving system, soon the staff there would have been idle, gazing at customers wishing to be served, and they would, in turn put even more pressure on the burger frying staff to release more burgers.

These unhealthy tendencies of optimizing areas before and after the bottleneck, without sorting out the challenges that have caused the bottleneck, are referred to as local optimization. Local optimization often leads to suboptimization, but it can be avoided if only people are keen to identify and address bottlenecks.

Anyone Can Identify A Bottleneck

Just as it is everyone's responsibility to ensure the system works optimally, it also helps if everyone takes responsibility in identifying bottlenecks and taking remedial action. For instance, in organizations where staff has internalized the aspect of working in systems, a cashier or any other staff member at the food outlet might not have waited to be prompted by the manager. Someone from a more relaxed section might just have moved in to lend a hand to the overwhelmed burger frying staff and then resumed their usual duties when the system was back to normal.

2) Leverage

Leverage is the ability of an individual or organization to influence a system such that returns per unit of effort can be maximized. This is important if you keep in mind that all resources, inclusive of time and energy, are scarce. The idea in leverage, therefore, is to try and gain maximum Return on Investment (ROI) from any of the resources employed. One way an individual or other system can manage such maximization is to take resources at hand as being interchangeable. Just as you exchange your financial currencies when the exchange rate is favorable to you, so can you exchange the resources.

For example, some people use their willpower to generate much needed motivation, while others use the same willpower to automate their systems in areas where it works best in a bid to make them stronger. If you utilize your leverage well, the Pareto Principle, that is also known as the 80/20 rule, will apply to your system. This means that you will be able to accomplish 80 percent of the work by making use of only 20 percent of resources that other people ordinarily use. With such efficiency, scarcity of resources will be the least of your worries.

How to Identify the Best Leverage Point

If, after deciding to switch to systems thinking and completing your planning, you manage to identify any major problems that are bound to affect the efficiency of the system, this is a great accomplishment. However, even if you are lucky in that the problems are solvable, you still have the problem of finding the right leverage point to effect changes to the system.

One thing you need to realize about effecting system changes is that, if you have a problematic system, and then you effect some change, it's pleasing to see the system working well, even without knowing how sustainable that smoothness of operation will last. Often it lasts only for a while before the problem recurs.

On the contrary, if your system seems to be working well, but, in your view, could do much better, and then you decide to effect some change, the entire workforce is likely to be up in arms. They will not understand why you are interrupting or modifying a system that does not appear broken or weak. As a competent systems thinker, if you have done your due diligence, you need not falter, because you will, very likely, have identified a suitable change strategy for your system.

In fact, what some people witness while trying to implement a solid systems-thinking strategy is that other people sometimes think you are not in your normal senses and may not be very cooperative. As discussed later, the management team involved in implementing systems changes at Griffin Health Services Corporation in Connecticut, USA, can attest to that. Of course, there are good reasons why people encounter extra difficulties when trying to implement systems thinking as opposed to other changes, and why there is dissonance in the whole process, and they are related to the ways systems respond to behavioral change.

Systems' Resistance to Change and Best Solutions

- Systems generally resist change

Just like human beings, systems are generally not very receptive to change. Any attempts to alter the conventional behavior within an organization are normally met with resistance, and sometimes there is even acrimony against the person or team trying to implement such change.

In fact, the more you express your determination to effect change, the more the system is motivated to put up resistance.

- Identify the right leverage point

Difficult as it is to implement change through systems thinking, it is possible to do it successfully if you manage to identify the right leverage point. Remember, the leverage point is that place within the system where, if you make any adjustment to it, the entire system will experience a big change. A complex system such as a city, a corporation or a country's economy can have more than one leverage point.

- Leverage points are often far from the visible or identifiable problem

This means that the location where the problem is identified does not necessarily constitute the leverage point. The same case applies to the timing. Just because a problem has been identified this week does not denote the time the problem began or the hour when the leverage point began to require attention. In short, what you see as the leverage point may not be reality. These are some of the situations where critical thinking comes in very handy because the system requires comprehensive analysis and synthesis before an effective and sustainable solution can be implemented.

- The impact of utilizing leverage points often begins with discomfort

One reason why people tend to resist change, particularly when it comes in the form of leverage points, is that its impact initially causes discomfort within the system, and the positive outcome only shows up in due course. That is why you need to be confident in what you are doing and to prepare your team psychologically if you are to work through leverage points. Weak management can easily be derailed by people's negative responses when trying to implement changes, but everyone is happy when the rough patch is over and the organization is reaping sustainable rewards.

You need also to be wary when you try to effect change through leverage points and the fruits show up almost instantly. Usually in such cases, the results are not sustainable and only last for a short period before the system begins declining once more. Obviously, in systems thinking the intention is to work towards sustainable efficiency.

- Interventions might backfire

Sometimes the changes you make at the places you believe to be the leverage points might affect some areas of the main system badly. Some subsystems might respond in a strange way while others may be practically hostile. Yet, for any changes to be good for the system they should not introduce new complications but solve existing problems and increase overall system efficiency.

As a good systems thinker, it is important to distinguish between the instances when you have gotten your leverage points wrong and when the system is experiencing the so-called calm before the storm. You need to learn to anticipate complaints and disbelief from the system community whenever you have gotten your leverage points right, because the changes you implement at those points have a strong impact on all areas of the system; and that is as it should be.

A significant number of stakeholders might pressure management to bring the changes to a halt even before they have been given a chance to produce the expected positive and sustainable results.

Examples of Working System Interventions
At Griffin Hospital, even when there was general scarcity of nursing staff, the institution created an additional level or workers to help in health care. These employees would not only perform the usual duties

of a nursing aide but also attend to patients' personal needs such as bathing and feeding, as well as maintaining room cleanliness.

Payment for these health care workers was increased above that of ordinary nursing aides to cater for the added responsibilities. Initially, the nurse aides already employed by the institution were trained appropriately so that they could fit in the new role, but later more staff was hired from outside the hospital. The goal of the hospital in creating the new level of health-care workers was to ensure the workforce provided better support to patients and coordinated inpatient activities more efficiently.

One might wish to know how the institution itself as a system expected to benefit. The answer is:

- Engagement of fewer employees

It was envisioned that since the nurse aides, with an improved pay package, would take over the services of tray delivery and patient cleanliness, the institution's need to engage more health workers to deliver food and perform housekeeping duties would be diminished.

- Improvement of patient satisfaction

It was anticipated that patients would appreciate the enhanced personal services and would be more satisfied and happy.

Generally, the new arrangement was embraced by physicians and nurses as well as patients, although there were some murmurs of complaints from employees who had previously worked as nurse aides. In short, the program was practically implemented, and it was hoped it would improve service delivery and cost the hospital less. Did that happen?

Well, a year down the line, the little murmurs of complaints had turned into audible complaints, and nurses, particularly, had become unsupportive of the program. In fact, there were no signs of a drop in the number of dietary and housekeeping staff. So, it seemed like the program had not helped the hospital cut staff costs. The discernible reality was that nurse aides' pay had been raised and they had received a new title, but when it came to the services they provided, they did not appear to have changed, and patients' satisfaction was not any higher. What was the problem?

The hospital had wrongly identified the leverage point, and that got the system pushing back. It was possible management had identified a position close to the real problem but not the exact issue. They had some apparent time constraints and did not, therefore, take sufficient time to identify the correct leverage point. This is representative of a phenomenon known as "better before worse."

In this case, the new strategy of nurse aides' service delivery had been generally embraced without opposition at its implementation, but in the long run it did not work for the hospital and everyone else in it. All the stakeholders, including the hospital management, ended up being unhappy about the system, and the new actions did not achieve their intended goals. This serves as a reminder that having changes unanimously accepted is not a sign they are the best for the system and everyone involved.

In the same vein, facing resistance, even stiff resistance, when trying to introduce changes in an institution, or any system for that matter, does not mean the changes are bad or unwarranted. Instead of bending to pressure to abandon the course of change, you need to be persistent and do what you can to convince everyone concerned to be patient. If you can manage to buy yourself some time, your changes will bear fruit and, down the line, you will be vindicated. If you are in top management, you probably only need to put your foot down and wait for the positive impact of your changes to show up later.

What should help you withstand inside and outside pressure is the knowledge that if you have correctly identified the leverage point, the system will sustain the newly-reached efficiency level.

Points to Help in Leverage

a) Adjust the rules

This means you need to modify the rules of operation, and that will help you realize the actions that are practical and can be performed within your system. This effectively means changing your daily habits and building solid system blocks.

b) Enhance self-organization

This means putting measures in place that will ensure natural, continuous improvement of the systems. Effective measures include getting staff to commit to working in a certain way or fulfilling specific goals, establishing artificial checkpoints and even deliberately putting up environmental constraints to keep everyone within the required working path.

c) Enhance the flow of information

This can be achieved by taking frequent measurements of progress made and doing this as objectively and accurately as possible. The information acquired from such measurements is then used to enhance the system by creating feedback loops.

3) Feedback loop

A feedback loop is a type of systems-thinking model that enables a system to learn how it is performing as compared to the expected performance. In order to achieve this, there is constant feedback where information flows into and out of the system.

The reason feedback loops are used is because this relationship that exists between the measurement and the item or action being measured is not linear but circular. When a measurement is taken, the system responds by bringing the quantity that has been measured nearer to the set goal, and that in turn causes the measurement to change.

The two kinds of feedback loops often used are:

(i) Balancing feedback loop

(ii) Reinforcing feedback loop

Chapter 11: The Balancing and Reinforcing Feedback Loops

Between the two kinds of feedback loops, the balancing feedback loop is the one most often used to maintain equilibrium or a stable status quo.

Balancing Feedback Loop

A good example of a balancing feedback loop is the thermostat. Keep in mind that a thermostat is relied upon to maintain a desired temperature level within a room.

Within the human body, homeostasis serves as the balancing feedback loop, and without it human beings would not survive. Although homeostasis plays an important role in the body, it is also responsible for making the body resist change. Yet behavioral patterns can only be modified for the better through accepting change. So homeostasis is a survival mechanism with a somewhat contradictory side.

Balancing feedback loops can be viewed as circles representing cause and effect, countering change by pushing in the opposite direction. They are sometimes considered to be negative feedback loops. With these loops, the more a push is exerted, the more the system pushes back, and that action and counteraction bring about stability within a system, sometimes viewed as stubbornness.

<u>Examples of Balancing Loops in Daily Life</u>

1) In biology

Consider times when your body temperature has risen, whether from illness, the environment or other reasons. The respiratory system responds by initiating sweating, and as the sweat evaporates, leaving the warm body surface, your entire body system feels cool. Hence, a balancing process is completed.

When the system has cooled down, the body no longer sweats as much, and there is not much sweat requiring evaporation. As such, there is less heat being drained from the body, and that means the body can now accumulate more heat. Thus, your body temperature begins to rise again.

When the converse happens and your body temperature drastically drops, you begin to shiver, and that shivering process ignites heat from the body to keep your system warm. That process is a balancing move geared towards reducing the initial temperature drop. That is why homeostasis is considered a good example of balancing feedback. In fact, it is only one among many biological functions that cells in human beings, as well as in other living organisms, use as balance feedback to keep their respective environments stable.

2) In the ecosystem

Consider the wild, for example, where there are animals of all kinds, some of them predators. Yet those animals have coexisted for thousands of years in their different sizes, strengths and feeding habits. When predators increase, they feed on their prey animals so much that their source of food begins to become scarce. Without enough prey to feed on, predators cannot thrive, and soon their numbers begin to fall. By the time the decline of predators becomes noticeably great, the animals that usually fall prey to these predators will have begun to increase in number, after a period of reduced threat. The whole process keeps the predator-prey ecosystem well balanced. According to ecologists, this self-regulating environment can safely be left to determine the level of predator existence that is sustainable; in other words, the carrying capacity.

3) In politics

(i) Regulating political parties' tenure

Balancing loops work almost in all systems, including political systems. For example, when one political party becomes powerful, whether through popularity or manipulation of sorts, it is likely to win elections. Forthwith, it is accorded more responsibility, including that of solving economic, health and other problems affecting the people. Since it is rare, if ever, that all problems can be alleviated in the party's span in power, the party is forced to take responsibility for all the problems ailing the country.

It is on that basis that many leaders from the governing party find it difficult to win their elective seats again in real democracies. The party, too, finds it an uphill task to produce another head of state after its time in office, because people associate them with the country's existing problems. This balancing loop is evident in a democratic country like the US, where no single party stays in power for too long, and the number of House representatives from the two dominant political parties keeps fluctuating almost evenly.

(ii) Regulating number of immigrants

Whenever there is an influx of immigrants crossing the US border, for example, US residents living along the border become concerned, viewing them as a potential threat as far as availability of high-paying jobs is concerned. They think they are the reason employers are likely to cut wages even as other fringe benefits are reduced. When such residents raise complaints with their local and federal authorities, the country often moves to make immigration laws stricter, so that fewer immigrants are able to come in.

Of course, such measures often cause certain industries to suffer due to continually rising employee wages, and investors threaten to move out of the country. When that happens, the same government reviews immigration laws, introducing clauses that allow certain categories of people to come into the US as employees, and that helps to neutralize the rise in salaries.

4) In social life

A balancing feedback loop often has a kind of implicit goal, and sometimes that goal is explicit. Take the example of two cars, yours and another one ahead of you. When do you normally increase your speed? It is when the car ahead of you also increases its speed. Conversely, you will normally find yourself reducing the speed of your car when the car ahead of you slows down. Can you see what is happening here?

Without stating it, you are trying to maintain a specific distance between your car and the one ahead of you, and the distance you are trying to maintain is predetermined and recommended. You are already aware of that distance, and that is why you accelerate when it is exceeded and decelerate when it is infringed. Sometimes you will introduce the balancing act yourself by stepping on the car brakes, but other times your car will be slowed by the road friction on a particular stretch.

5) In the economy

(i) Controlling a monopoly

Observe a company that began small, and then, due to the monopoly of its products or services, it grew rapidly and became really big. Does it remain that way forever? Hardly—normally the company capitalizes on its monopoly status to set exorbitant prices, and it continues to reap super-normal profits. Eventually the hue and cry from the public becomes so loud that the government pays attention. Soon the government introduces regulations intended to bring prices down while also quelling public noise. Some of the regulations of this kind are geared towards opening space for other investors to enter the market and break the monopoly. Others come in the form of heavy taxation targeting revenues of a certain level, and these can force the company to break up and reinvest as different, smaller outfits.

(ii) Inventory management

When a company realizes it has too much stock built up, and probably the cost of storage is not worthwhile, usually it decides to cut the prices of those products whose stock is high, in a bid to get more customers interested in buying. That is when people see unprecedented price cuts in sales offers, and if the stocks are really high, the low prices are retained indefinitely.

What happens to the stock as the number of regular product customers increases? Obviously, soon the stock is depleted. The pricing of new stock is then bound to be higher than the earlier lowered price.

Implication of Balancing Feedback Interventions

When an organization goes ahead to install or implement balancing interventions, the system is seen to respond when there is a deviation of performance from the initial goal. The pressure for change that existed within the system also begins to weaken when it is apparent performance has begun to improve, as long as the actions towards performance improvement are being driven by balancing feedback. It is important to know that there is a possibility for the elements in the balancing feedback loop to break down and deter the system from producing the desired results or reaching the anticipated goal.

With this awareness, there are always interventions that can be put in place to ensure the success of the process.

1) Identify the process limit

For one, it is important to identify the limitation of the balancing process or even the expected goal. It is true systems have set goals, but many of the processes happen to be designed to attain different goals, such as pressure relief. As such, it is important to adjust the goal of the balancing process accordingly. It is also helpful when the goal is adjusted due to changing aspirations. In case the goal structure is itself limiting, it is advisable to find means to remove or even reduce the constraint in order to ensure performance rises to a higher level.

2) Get rid of delays, or at least reduce them

It is recommended that the time taken to measure system performance be reduced so that it takes less time for the corrective action effect to be felt. The reason for this is to avoid excessive control, which is exercised when there are long delays. A good example is the time it takes to detect the temperature in an old-fashioned shower. Obviously, a lot of heating will have taken place before the change in water temperature is felt.

3) Evaluate the corrective action in advance

Sometimes people want to implement what they term corrective action only because they are familiar with that actions or process, or because it happens to fall under their area of specialization. However, unless those actions are suitable for the current system process, they will not improve the system for the better. If they prove to be inappropriate, the organization will have incurred unnecessary costs of implementation and unwarranted hassle, and if the corrective action makes no impact at all, the hassle of change will still have been unnecessary. That is why any proposed actions need to be critically evaluated to ensure there is good reason to believe they will truly improve the system.

Balancing Feedback Cycles

Examples Associated with Service Delivery

In the example that follows, the balancing feedback could be applicable to supermarket queues or those at any cash register, bank, gas station or even at hospital emergency rooms.

Where customers, or patients, in the case of medical institutions, come in and are attended fast so that waiting time is impressively short, they go home happy and spread the word that the quality of service at those stations is exemplary. Of course, everyone wants to be served at a place where quality of service is high, so those who can will flock to those stations. That results in higher customer traffic in those particular stations that have recently become popular.

Is it surprising that the stations now have long queues? With longer queues, the average time it takes to serve one customer becomes longer, and the customers conclude that service at the station has dropped with time. Because of that belief, some of the customers will, very likely, not make return visits henceforth, and they will seek a similar service elsewhere. And, just as word of mouth spread that quality of service was high in one establishment, so will it spread that quality of service has declined, and potential customers will divert elsewhere.

Consequently, there will be fewer customers coming in, and so the average waiting period for one customer will drastically fall. Again, everyone being served will begin talking about the elevated quality of service. Thus, the cycle will continue till a good and sustainable balance is attained.

Example Associated with Training

When staff competency is on the decline, or when the level of skilled manpower has dropped, system performance drops as well, and it is clear the organization is not doing as well as expected. That position elevates the push for staff training, and that leads to a rise in skilled manpower within the organization. When it is clear the trained workforce and the organization are performing well, management stops putting emphasis on staff training. After a couple of years, the relegation of staff training to the backburner begins to show in poor system performance. Very likely, staff will be using outmoded ways of working and solving problems, and the organization will have been left behind by others in the industry whose staff will have kept abreast of changing trends.

Reinforcing Feedback Loop

As for the reinforcing feedback loop, it is responsible for introducing and enhancing either growth or decay within a system. When one introduces change towards a given direction, the corresponding results are observed in direction at even greater magnitude. Effectively, without being well controlled, a reinforcing feedback loop can result in either a virtuous cycle, denoted as positive (+) or a vicious cycle, denoted as negative (-). This will eventually lead to the value of the system either soaring to infinity or declining to zero in the respective scenarios.

Two good examples of a virtuous cycle and a vicious one are marketing and global warming, respectively.

Image: A Stampede Representing A Feedback Loop

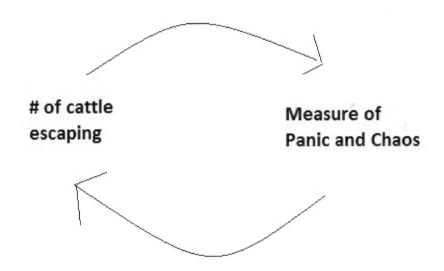

Installation of a Feedback Loop

The reason organizations may wish to install feedback loops is to ensure any desired behavior within the system is kept consistent, and that, when there is some failure in the consistency of behavior, the damage occasioned is kept in check. In short, the feedback loops act as firebreaks or checkpoints that are always at the ready to provide support when a new plan is threatened or to provide appropriate evaluation for such attacks.

When it comes to installation of feedback loops, everyone needs to be vigilant and alert to how the balancing feedback around is working. There are counterbalancing forces within our bodies and in our environments, and they are also present in relationships and within our respective cultures; the negative thing about them is that they tend to attract us in the direction of undesired equilibrium.

Fortunately, it is possible to overcome such unfavorable states by making some serious adjustments, creating new defaults and baselines as well as fresh social norms.

The main reason for reinforcing feedback loops is to optimize the processes that are in place, and to ensure we set for ourselves a bar that is progressively higher as time goes. In short, long-term growth needs to come incrementally through daily improvements, and today's improvements should be built from yesterday's improvements. It is the accumulation of daily improvements that progressively constitute long-term growth.

In a nutshell, the three mental models just explained—the bottlenecks, leverage and feedback loops—combine to make up the best multi-tool for an organization's growth.

Measurements are taken and self-awareness enhanced in order to identify any challenges that may develop into constraints, and behavioral experiments are conducted with the aim of validating any assumptions made. You know you have correctly identified the bottleneck if the single change you make to an assumed obstacle or challenge ends up changing everything about the organization or system.

It is then that you redirect any extra time you have, as well as time and energy, towards working on the bottleneck. You try to achieve the greatest leverage interventions at your disposal in a sequential order. Since your goal is to have a smooth working system, once you realize the challenge you had identified is no longer inhibiting the smooth running of the system, you need to cease directing your energies toward it and begin afresh to check the process and see if you can identify another bottleneck. Until your organization begins to perform the best it can, you are likely to identify continual bottlenecks in the system if you look keenly.

That continuity of identifying bottlenecks and clearing them creates momentum that keeps the system working like clockwork, pushing the human resources and collective systems to a higher level every time, with every iteration. Hence, emerging bottlenecks become weaker and less constraining, meaning the system will gradually approach its optimal level of performance, all this courtesy of the feedback loop.

The entire process involves identifying the bottleneck within the system, performing various experiments to remove the bottleneck and, once the bottleneck is cleared, repeating the process again and repeating it to ensure sustained high performance.

Reinforcing Loop Examples

It is important to remember that, in a reinforcing loop, one action produces results that end up influencing more of that action, and in so doing, it causes either system growth or decline. Since the reinforcing loop is one of the most important structures of systems thinking, it is important that we cite some practical examples to show how it works in everyday life.

Reinforcing Loop In the Finance Sector

Take the example of a bank savings account. The way the reinforcing loop works is that the account begins with an amount of savings that constitutes the principal amount. Over time, there is interaction between the principal and the interest rate, causing the principal to bear interest. Over the financial period that follows, the principal sum will be higher because the interest already earned will have been added to the original amount that made up the principal. As such, when next the principal interacts with the interest rate, which, in this example, is assumed to be constant, the interest earned will be higher. This time around, the amount of interest added to the grown principal amount will be even higher than it was in the previous financial period.

Owing to the manner in which this system structure has kept reinforcing itself, it has resulted in exponential growth. Under different circumstances, the same system might have resulted in exponential decline. Sometimes it takes a while for the exponential change to become noticeable, and it only gets noticed when it reaches a particular threshold. When that point is reached, the system structure appears to change so fast that the people involved with it wonder how the great change suddenly happened. In short, they often do not realize that the change was not sudden at all, but gradual. But before, the change wasn't substantial enough for them to have noticed it.

Best Strategies to Work with Reinforcing Loops

(i)   Work on sustaining the growth

What do people do about the system when it is working well? In many cases, they let the system be. As long as there are reinforcing system structures producing favorable results, what is referred to as a virtuous cycle, people tend to ignore the system. After all, the cycle of good results is evidently continuing. However, such an attitude is detrimental to the system as unattended growth cannot sustain itself indefinitely. The exponential growth is likely to slow down after sometime, and probably stop entirely at some point if there is no intervention. Ideally, it is at the height of great performance that you need to seek means of sustaining the exponential growth.

(ii)   Break the cycle of decline

What happens when the result of the reinforcing structure is a decline in system performance? At times when the results being produced by the reinforcing structure are unfavorable, you need to take remedial action as soon as possible. The best step to take is to aim at breaking one of the existing feedback loops, because with one feedback loop within the system interfered with, the cycle cannot continue as before. In short, the cycle that was producing undesirable results cannot reinforce itself after you have broken one of its sets of feedback loops.

In an organization, once you have done away with the system's self-reinforcing weaknesses in this manner, you have room to create strong structures geared towards producing positive exponential growth.

(iii) Identify the growth limit

Often, management celebrates good performance when it happens and continues, but rarely do they think this positive performance has its limit. The best advice in a case where the organization is going through positive exponential growth is not to rest on its laurels. Instead, it needs to proactively try and find out what the systems growth limit is under the circumstances and within the economic environment. It is always better to anticipate a halt to growth than to be caught off guard.

When it comes to limitation in growth, you need to realize it comes after a period of success. So the idea is to avoid being blinded by success and keep reveling in it as though it is there to last a lifetime. There is actually an action emanating from the balancing loop that brings the unattended success of a reinforcing loop to an end.

What happens in a system experiencing exponential growth is that the successful actions within the system, with its reinforcing loop, lead to desirable results, and those results further enhance the successful actions. However, as the reinforcing loop continues its cycle, it comes across and interacts with some limiting factor. That interaction leads to an action or actions with a slowing-down effect.

In effect, therefore, the action with the slowing-down effect subtracts some goodness from the results, and the exponential growth no longer continues at the same pace. The system performance begins to, inevitably, slow. In business organizations, political parties and other areas, the reinforcing loop generally operates without any apparent hitch, and no one can imagine the presence of a limiting factor. Only when the system performance reaches a certain level does the limiting activity begin to have a negative effect, basically slowing the growth that had once continued unabated.

As such, it is not surprising that it becomes confusing to management and everyone else involved when the growth the organization had experienced for some time begins to significantly slow down. Unfortunately, since most people do not realize there is something like a limiting action, they continue to put emphasis on the actions known to produce positive results. Needless to say, when the system results fail to pick up despite concerted efforts, there is even more confusion among the stakeholders. Unless the system gets people who understand the aspect of the limiting factor within systems thinking, the negative results caused by the limiting action will continue to cause further decline to the organization's results.

(iv) Look for a potential limiting factor

The best defense is a good offense. This means you should not wait for the problem to surprise you, but instead proactively check the system to see if any action being undertaken is likely to become a limiting factor in the future. One fact you need to acknowledge is that nothing is created from nothing, and so when there are good results showing somewhere in the system, very likely, there is another area of the system being adversely affected. In short, every good action, in this case performance growth, comes at a cost, and that cost needs to be acknowledged so that it can be controlled before it can have a significant negative impact on the system's overall desired results.

(v) Check if the structure supports success to the successful

Sometimes limiting factors do not just automatically emerge, but we, unknowingly, create them ourselves. Company management does the same as well, without being conscious of it, and the trend is prevalent in many places including in the world of sports.

Example of Resource Allocation in Sports

Take, for example, a college that excels in two major sports, basketball and soccer. If the college management begins to reduce the resources allocated to soccer and begins to increase the allocation for the basketball team, inevitably there are some activities the soccer team used to do that are now going to have to stop or lessen. The coach for the soccer team might have to book the team into cheaper hotels when they are out for training or at competitions. What does that do to the soccer players psychologically?

It is bound to demoralize them, as they will assume they are less valued by the college as players. They may even begin training for shorter periods because the cheap hotel might be very far from the training ground. In the meantime, the basketball coach may have improved accommodation conditions for his team and even be afforded more training sessions and higher quality gym facilities. Is the basketball team not likely to perform much better than before?

Incidentally, it may not dawn on the college management that the reason the soccer team is performing dismally, as the basketball team excels, is because resource allocation is favoring the basketball team. As a result of failing to identify the limiting factor, management is likely to be excited about the continued success of the basketball team and will continue to shower them with more resources.

This is a mode of reinforcing structure referred to as "success to the successful," and it involves sustaining a self-fulfilling prophecy. In the case of the two teams, management wanted to see the basketball team succeed and they therefore gave them more funds than before while reducing those of the soccer team, and there the cycle of unbalanced performance took root. In short, it is important for management to analyze the way they support different departments and individuals and see if they have played a role in the high performance of some while catalyzing a decline in others.

Chapter 12: The Process of Switching to Systems Thinking

People have been using linear thinking for a long time. However, many have found it preferable to switch to systems thinking, and not because it is the in-thing but because it seems to produce better performance results in business organizations, social systems and elsewhere.

Does this mean that linear thinking should be done away with? No. There may be situations where linear thinking will come in handy, but, given that systems thinking appreciates that every system has a way of affecting other systems, it seems to make great achievements sustainable for longer periods than linear thinking, hence gaining currency among successful bodies.

We can use the analogy of physics Nobel Prize winner, Albert Einstein, who was at first interested in quantum physics, but soon realized he did not really like it as a working choice. He even began to question the credibility of quantum physics, but that was because he did not understand it. However, although he chose to base his scientific work on Newtonian Physics, he soon realized and acknowledged it did not provide the most answers to his questions. So, out of necessity, the great physicist returned to quantum physics, and needless to say, he became a master in it.

Does that mean Einstein rejected Newtonian Physics altogether? No. What it means is that he embraced both lines of physics, but when it came to solving problems and explaining his findings, he often found it easier and more convenient to apply quantum thinking as opposed to the Newtonian approach.

In the same way, although people have not entirely rejected linear thinking, they have not picked systems thinking at random. They have only discovered and acknowledged that there are many questions that come up when working and linear thinking does not provide the answers. It is important to appreciate that linear thinking works on the basis of cause-and-effect. In linear thinking, you should anticipate a single effect from one cause.

For example, if you have been driving for a while and your car tank runs dry, the car stops. The car running out of gas is the cause while the stopping of the car is the effect. It is as direct as that. These are the kinds of problems that linear thinking is good at solving, but when it comes to more complex problems, only systems thinking can help.

As you adopt systems thinking in your organization, you need to be prepared to study the system properly, so that you can distinguish a problem from a symptom. Many organizations have spent time and resources designing and implementing interventions, only for the system to continue performing as poorly as it did before, and, in some cases, worse.

Interventions that bring short-term solutions to a problem are just like Band-Aids that stop the bleeding and sometimes numb the pain but do not heal the injured part of the body. In fact, if temporary interventions are not correctly classified and treated as such, there is a risk of the system problem quietly becoming more serious than it initially was. Worse still, since the exacerbation happens underneath, people go about their duties thinking all is well, and by the time the brewing problem is noticed more resources will be required to solve it and the action required will, very likely, be drastic.

A Socio-Economic Problem Erroneously Solved Linearly

Take the problem of mosquitoes, for example, which for many years were considered a health and social menace. The people in charge of public health in many countries used the pesticide, DDT, to fumigate public grounds in order to stop the mosquitoes from breeding. That worked well as far as fighting the mosquito menace was concerned. For example, there were fewer people visiting hospitals because of malaria, and in less developed countries, where infrastructure was still very poor, fewer malaria-related deaths were reported.

Did that make the world any better? It actually did, until there was a hue and cry in the US over the side effects of the pesticide. One activist, Rachel Carson, was particularly vocal on the negative impact the chemical was having on birds, saying its use led to reduced bird populations. When the conservative wing of politicians joined the fray, there was enough pressure for the government to ban the pesticide. Forthwith, the use of DDT to kill pests on farms was discontinued, and its use for malaria control was discontinued too.

What happened then? The malaria menace was back. Of course, other ways have been invented to control mosquitoes and malaria today, but because the policy change in the case of DDT ban was not thought out in the context of whole systems, whatever action was taken—either the use of the pesticide or its discontinuation—ended up causing unwarranted problems.

One main reason systems thinking is preferable in organizations is that it introduces powerful working tools as well as enlightened perspectives to issues, working strategies, problem-solving and organizational diagnoses as well as leadership.

Comparison Between Linear and Systems Thinking

Linear Thinking	Systems Thinking
Organization is viewed as an assembly of different departments, sections or functions.	Organization is viewed as one system.
Interest is in the content.	Interest is in the processes.
Often, it is symptoms that are treated.	Interest is in the underlying dynamics.
It is common to assign blame.	Often, people try to identify behavior patterns.
In a bid to create order, focus is placed on controlling chaos.	Focus is placed on identifying behavior patterns within existing chaos.
People concentrate on the communication content.	Although attention is paid to content, people are also keen on studying interactions as well as communication patterns.
Belief is held that organizations are orderly and that it is possible to predict their future.	Belief is held that organizations are unpredictable and that they exist in a chaotic environment.

If you want to become a systems thinker, you have got to change your mindset from what it was in linear thinking, which was to view things in terms of cause followed directly by the effect, to one where there are different angles to one thing. Obviously, things are easy in linear thinking as there are no complexities involved, so for you to succeed in becoming a successful systems thinker you need to be patient and persistent. On top of that, you need to be curious.

If the organization where you are working has been using the linear approach, expect to feel somewhat isolated, even if you happen to be in management. If you are the only one trying to think in systems, others will find your way of reasoning odd, and that is because they will not understand you. It can cause you mental turmoil, of course, trying to accomplish daily chores with your colleagues with a linear mindset while you are trying to exercise systems thinking and to establish possible bottlenecks. Some may look at you as if you are discussing an organization alien to them and will often give you a look of confusion, but that should not deter you from pursuing what you know to be fitting for the system.

One reason you need to exercise patience while still being persistent is that, while you are aware of the existence of both the linear and systems-thinking approaches, many of your colleagues will have been exposed to only the linear approach. They may not even appreciate that the environment within which they work on a daily basis is a system. So their behavior towards you is quite understandable.

Easy Tips to Systems Thinking

People who have worked in organizations using the linear approach for most of their lives—and such organizations are very many, if not most—they need to be led and shown the easiest way to adjust to the systems-thinking approach. Here are some helpful tips:

- When you notice someone doing things that are not exactly helpful, instead of blaming that person, try and understand, or simply ask what it is that influences the person in taking those actions.

- Instead of declaring that you know the answer, offer an additional perspective on the issue.

- Instead of paying attention to solely one item, be prepared to evaluate all variables involved; all that affect that particular item you are handling.

- Instead of considering the content of people's opinions, analyze the process they take to convey the content, how they frame what they say, what they are holding back instead of saying it and what the common theme in different people's content is.

- Instead of dwelling on any negative sentiments and behavior, try and establish that which is motivating people to exhibit their various behaviors; in the process, try to learn if there is a bigger underlying problem.

- Instead of focusing on what people are doing, consider the dynamics of the organization, with a view toward understanding the forces that are pushing people into behaving in their various ways.

Learning to Distinguish Problem from Symptom

If you are serious about making a shift from the linear approach to thinking in systems, you need to learn to distinguish a problem from a problem symptom. If you do not, you risk investing yourself in alleviating an apparent problem only to realize you have wasted your time and resources and left the problem intact. Of course, if you are a project manager and you spend massive resources only to have the project still suffering from its initial problem, all will not be well between you and the organization's top management.

Where linear thinking is employed, people tend to focus mainly on addressing the behaviors that are explicit, and sometimes those behaviors are not the real problems but mere problem symptoms. As has been mentioned, alleviating problem symptoms does not eliminate the underlying problems. In fact, there are instances when getting rid of the symptoms increases the magnitude of the underlying problem. Actions may be taken in the linear approach that eliminate signs of a problem in one area of the organization, while the same actions cause a problem in another area of the organization.

However, if the manager in charge works by thinking in systems, it will occur to him to look at all the sectors of the organization that have a direct or indirect link to the area that appears problematic, and then proceed to determine the real cause of the problem while analyzing the possible repercussions of any possible solutions. If the real problem is identified and dealt with, chances are that all symptoms associated with that problem will disappear.

Clues That What You Have is Only a Problem Symptom

Since it is imperative that you address the real problem if the system changes are to bear real, sustainable benefits, it is important to know whether you are looking at a real problem or mere symptoms of it.

You know you are yet to identify the problem:

1) When the discussion around the issue is too much compared to the magnitude of the apparent problem.

If, for instance, after relocating to a new site, people in the office keep complaining about the color of the doormat at every opportune moment, you need to translate that as being a symptom of an underlying problem. Probably the people are fed up by the management's style of making major changes without consulting them and only issuing instructions at the time of effecting changes. Their complaints could be a sign that they require an audience with management to vent their anger over what they view as being taken for granted.

If this is the case, no matter the color of doormat, the complaints will persist or shift to something else, like, for example, the design of the office tea mugs.

2) When people leave the problem to persist even when they can solve it

Why would people keep complaining about a problem that they are capable of solving without outside intervention? If you look at the thing being complained about and in your fair judgment anyone could have solved it if they wanted, then it is time for you to delve deeper into the issue or to study the environment to see if you can pinpoint what the real problem is.

Say, for instance, you are a tutor and you enter the classroom only to find the students complaining about the stuffiness in the room. On looking around you realize none of the windows are open. Would you really consider the stuffiness of the room to be the real problem or a symptom?

Obviously, if the students cared to open the windows, or even some of the windows, which they could if they wanted, the room would not have been stuffy. Probably they were upset some pit latrines had been dug very close to the classroom, and in their opinion the stench would fill the classroom if they opened the windows. The pit latrines might even have been too new to have had an impact on the environment at the time.

The problem might also have been different. Probably the students had always wanted the school to employ toilet cleaners but the school insisted on students cleaning their own toilets, which they could only do in the morning as per the duty roster. The students' complaint about the classroom stuffiness was a sign of discontent between the students and the administration, and even if the teacher got air-conditioning equipment installed in the classroom, that discontent would still remain.

3) When the problem persists

How many lives can problems have? If every year you are spending an amount of money to solve a recurrent problem, then it is really not the problem but a symptom of a real problem. As such, you need to commit to looking for the real problem within the system, or from systems linked to your system, and address it.

Even if that problem goes away, but reemerges in another form, avoid labeling it a new problem, because if you do and spend time and resources solving it, it will still morph into another form and reappear as a problem.

Say, for instance, in one week 20 students report at the school's sickbay with malaria. Would you call malaria the problem? Can you be certain you will not have more students reporting with malaria in the coming days, weeks or months? It will help to find out what it is that has made so many students contract malaria at the same time. If you analyze the situation well, you may establish that the area around the students' dorms is unkempt and has long grass, and it happens to be the wet season.

In short, what you have may be a management problem, requiring recruitment of staff to maintain the grounds, or the problem of bringing in a local authority to fumigate the areas around the school or college. You can still invest in mosquito nets for your students as an alternative or complementary solution.

4) People are emotionally restricted

There are some things that people fail to address just because nobody else has ever spoken about them. As such, even if it will help the organization to do those things, nobody ever gives them a chance. Sometimes people just laugh off when someone who wants to be proactive makes a related suggestion, seeing the suggestion made as farfetched or unachievable. In fact, there are many instances when solutions are implemented in an organization which could have been implemented years before if only one imaginative person had received the backing of colleagues and management. Such restrictions that make people afraid to try things are referred to as emotional barriers.

You could, for example, have a workforce that seems robotic and often appears bored and uninterested in anything beyond its routine work. Suppose one of the newest department managers suggests sponsoring employees for a staff dinner in a good restaurant on a quarterly basis. Some of the old managers might be horrified by the idea. What will they discuss with their juniors outside of work? Isn't that too big a privilege for employees—something that would be better extended to senior management?

That was the mindset of sailors before the days of Christopher Columbus. Sailors feared sailing southwards beyond the equator, and not because of the presumed shape of the earth, as is sometimes said, but simply nobody else had sailed that far and reported back about what it was like. Emotional barriers like that one are a result of stunted imaginations. Systems that do well encourage all staff to be imaginative and to feel free to express their ideas.

In the example involving the suggestion to sponsor staff dinners once in a while, such a habit would help to break down unnecessary barriers, and back in the office employees would find it easier to articulate their problems to their supervisors and managers. That would then create an environment where problems can be discussed and solutions sought amicably, in a way that does not affect productivity.

Take a case where an employee needs to shop for her child before the kid resumes school the following week. Fearing the head of department might not understand the importance of the employee taking some time off, the employee might just call in sick one day in the week. In the meantime, what happens when employees, including their seniors, have free communication?

The employee would have communicated the need to her senior, who might have suggested that instead of taking the day off, the employee report to work as usual but catch a ride with the purchasing officer sometime in the day when he is going out to deliver purchase orders to suppliers. She would be dropped at whatever outlet she wanted to shop at and be picked up by the purchasing officer on his way back to the office. Here you would have one happy employee, with her need met so there's no stress at work, and the company would not have lost an employee's service for an entire day. In systems thinking, there is no room for stunted imagination. Both junior and senior members of staff need to think globally, as opposed to linearly, with emotional barriers restricting their thinking and actions.

5) You notice a pattern in the presumed problem

If you can, for instance, foretell when a problem is about to emerge, very likely what you are terming a problem is only a symptom. Take, for example, a student who always reports to school late after school holidays, and the excuse he gives is always that he had been admitted to hospital with respiratory problems. Is illness really the problem? Is his lateness the problem?

You may wish to inquire from the student where he goes on holiday, because the weather is probably not good for him. However, you may realize the student falls ill because of the nature of the work he does over the holidays. Probably he works two or three jobs, often leaving home before dawn and going to bed past midnight in order to earn money for school fees and daily upkeep. At the end of the day, because of your capacity to think in systems, you might realize the issue at hand is one of a financial nature, and it can be alleviated if you present the student's name to the administration so that he may be considered for financial assistance.

If you get the student's fees issue sorted out, the hospital admissions and school lateness that have formed a predictable pattern will become a thing of the past. A problem should not have a cycle or a pattern. It should not be predictable, otherwise it is not the real problem.

6) The presumed problem has been around for sometime

Does anyone want to live with a problem? If an organization has allowed a problem to exist year in year out, then, very likely, it is not really a real problem. It is probably something someone or some people in the organization uses as a scapegoat when they have not met their targets, or to win sympathy for their own known reasons.

Of course, it may be argued that nobody wants to live with a problem, and that is true, but sometimes people procrastinate solving a challenge because, unconsciously, they like the idea of discussing it or dealing with it. So, whenever they purport to solve the apparent problem, they don't solve it completely, so it hangs around often, to be cited as a departmental or organization problem. The point is, if nobody has found it necessary to spare time and resources to deal with a matter conclusively, then there is high probability it is not a problem at all, or it is a smokescreen.

7) When the workforce is under stress

In organizations where employees do not feel free to express themselves freely for fear of being victimized, usually where management is domineering over its employees, junior staff tend to have myriad complaints. They complain about almost anything, because theirs is a way of expressing dissatisfaction with the environment. As such, the real problem, which is an unhealthy working environment, remains concealed within unrelated complaints, which might range from too much work to too little tea. In fact, it is easy to tell that what you are listening to is not the real problem because, no sooner do you alleviate it than another complaint surfaces. They are even likely to be aired in a string. When thinking in systems, you get the opportunity to analyze the working conditions and all circumstances surrounding complaints, and soon you will learn there are underlying dynamics igniting the apparent problems.

Chapter 13: Challenges Encountered in Systems Thinking

Although thinking in systems is the way to go for organizations that want to scale their success to great heights and maintain it there, there are a number of challenges facing this working approach, before and during implementation.

First of all, although people accept the idea of systems thinking when it is explained to them that a system is a composition of smaller systems that support one another, many do not really appreciate that this description is different from an assembly of subordinate parts. And even for those who do, there is the challenge of identifying a tangible system; a problem of conceptualizing it.

People accept that all the sections of a system or the constituent subsystems are interconnected, and that if their performance is enhanced, the entire system will work better too. However, when it comes to identifying areas that require modifications or streamlining, many people tend to go for those parts that can be seen with the eyes. Fixing the physical parts of a system, in the case of a manufacturing company regularly operating machinery, does not necessarily constitute efficient system management.

Example of Deficient System Management

Take the case of a car salesman dealing with secondhand vehicles. If the firm he works for notices his sales have dropped, the management might rightly conclude there is something wrong somewhere.

However, since, as has been noted already, people tend to address issues that can be observed and easily assessed, management is likely to extract the salesman's sales records, probably for the last quarter, to confirm their observation that he is really doing badly. After all, even when they look at the inventory they can see the cars are still in the yard, waiting to be sold.

Since the salesman is not performing as expected, management is likely to think of ways of motivating him. They can raise his commission or his monthly salary. They may also decide to give him targets to meet, and since nobody likes to fail a test, it is expected he will work hard to achieve those preset goals.

Alternatively, management might tell him there will be ranking for all the company's salesmen, and it will be clear where he falls. Hopefully, wanting to be seen as a good performer amongst his peers, he will work hard towards improving his sales record. Management might even decide a threat is the best option and warn the salesman that if his sales do not improve, he will be fired.

Will management have been thinking in systems in this case? Will it have been addressing the real problem? Is it not possible that car sales have dropped because of problems that are not visible with the eyes? Many times, even when following the approach of thinking in systems, we have the shortcoming of ignoring things that we cannot see with our eyes or touch with our hands. In the case of the poorly performing salesman, if management did a comprehensive analysis of the system, it might find there are other inadequacies that are likely to have adversely influenced the dismal level of car sales.

Possible Invisible System Problems

- Lack of relevant sales training on the part of the salesman

- Inadequate or poor quality promotional materials

- No budget allocation for the salesman to pay for vehicle cleaning

- No authority to give buyers after sales warranties

- Poor location of the car yard, for example, next to a slum

The point is, management could be right in that the salesman does need be motivated, but if lack of motivation is not the problem, car sales will not increase despite better pay and other benefits.

166

Therefore, there is a need to do a comprehensive analysis of the situation before determining where the problem lies. It may be that the man is really eager to sell but the people who can afford to buy cars are afraid to venture into the neighborhood where the car yard is located.

Some Behavior That Mars Systems Thinking

There are several occasions when people acknowledge from what they learn that thinking in systems is helpful, but when it comes to practically adjusting to the approach, it becomes a problem. Often this happens because people find it hard to drop old habits of linear thinking. In fact, sometimes people are not even conscious that some of their own attitudes and behaviors are sabotaging the proper implementation of systems thinking.

In order to bring out the negative attitudes and behaviors that do not enhance systems thinking, but rather create problems, those negative aspects have been formed into statements that are easy to understand here. Of course, it is important to determine the steps to be taken before you can conclude the person is in linear-thinking mode, but these statements should, at least, put you on the alert. If you are the person saying these things, ensure they are coming out of your mouth out of sheer habit only. Your thinking should be in systems and you will proceed to analyze and try to solve problems globally, and not from an isolated area.

- This has to be fixed immediately!

Is the problem in the urgency? No. It is still possible to attend to a problem quickly while thinking in systems. However, when people speak of fixing things forthwith, they are often looking at the visible problem and taking it to be the actual problem or the bottleneck. Take the example of the employees complaining about the color of the doormat. You may look at the doormat and, at a glance, concur that the color is too dull for an elegant office. Considering the doormat was just a problem symptom, how many times do you think the doormat would be changed to satisfy the employees? Infinite!

In short, hastening to fix the presumed problem before ensuring you have a good grasp of the situation and have ascertained you are addressing the real problem is a recipe for failure. If you have been used to linear thinking where fixing problems as you notice them is common, you need to consciously remind yourself to halt and think critically of the apparent problem and its associated system areas.

- Do something about this part that is problematic

This is commonly referred to as the bandage mentality, where, for instance, replacing a valve along the water system enables the water to keep flowing. If you do not analyze the problem, the new valve you fix may only last a short while before the water problem recurs.

Probably all the valves you replace happen to be rusty. When you replace the valve and procrastinate on doing a thorough analysis of the situation, are you not losing the opportunity to find out if the water has, for whatever reason, changed its composition? Suppose the water is now too salty and that is the reason it is causing the metallic valves to become rusty so fast. Where is excess salt flowing in from? Suppose the water is not even fit for consumption. As you can see from these questions, going for a temporary solution hinders systems thinking and, hence, your ability to solve bigger, and often more sensitive, system problems.

It actually acts as a problem cover-up and that is dangerous as the real problem could brew to unfathomable proportions and affect several subsystems. In a company, a problem that initially affected one department might fester and affect other company departments, and before you know it the problem can only be solved with the involvement of the board of directors or even the shareholders.

- We must complete the budget before the fiscal year ends

Budgets are good and necessary in an organization. However, they tend to be given too important a place in linear thinking, especially when it comes to their timing. For that reason, management ends up using budget provisions that are hardly realistic, while waiting just a little longer for crucial feedback from constituent systems would have provided more realistic figures.

Sometimes this linear thinking leads to departments utilizing allocated funds only because, in their view, it is important to utilize that money within that particular fiscal year. That kind of expenditure might not be very helpful to the organization because, often, the action will have been taken before due diligence had been carried out. In fact, it might not contribute to the organization's profitability in the short-term or even in the long run. In government systems and other nonprofit organizations, actions taken with this attitude of beating deadlines are often not geared towards increasing the system's efficiency.

- It is important that we give an immediate response!

Why the anxiety? If you panic when you notice there is a problem somewhere, you are likely to go for linear solutions, reacting in a kneejerk manner. While speed in solving the problem may be of the essence, operating calmly is even more important, because only when you are calm can you appreciate the problem properly and find a solution through critical thinking.

- So what?

Have you noticed some people are indifferent to problems even when they are very much part of the system being affected? People who exhibit a carefree attitude to problems are not very helpful when it comes to improving the efficiency of the system. If you are going to succeed in thinking in systems, you have to be proactive and imaginative. You need to be prepared to explore possibilities, not only of the problem causes but also of feasible solutions. This encompasses an element of curiosity, a positive attitude towards adventure and even playfulness.

- More information is required here.

Okay—so we do nothing now? When people make statements like these when faced with a problem, it looks like they are evading responsibility or are afraid to take any action whatsoever. Can you not, at least, begin to lay out a strategy to study the problem area? Can you not begin to send your staff to make inquiries of other departments or stakeholders that might help shed light on the current unpleasant situation?

Making a statement that you need more information sounds like you anticipate that the information will give you additional power than you already have, which is not true. In systems thinking, people analyze a situation and do not wait for information to arrive; they look for it. In short, someone speaking in the manner reflected by this statement is, very likely, not thinking in systems but thinking linearly.

- Are you not overthinking?

No, I'm not. I'm just thinking broadly and trying to see the problem from different perspectives. Unfortunately, people who have used the linear approach to problem-solving all their life, and are conservative thinkers who want to follow the beaten path and deal with everything in a conventional way, find system thinkers overstretching their thinking. Often, when they allege you are overthinking, all they really mean is that you are thinking unconventionally or differently from them.

- Let us think of us first

This is a common mentality among heads of departments in schools, companies, government bodies and other organizations. They want their departments to be seen to be doing well, or better than the others, and that means they do not really care how the other departments, or ministries in the case of government, perform. That is the nature of linear thinking, where constituent departments of an institution compete to outdo one another as opposed to working in support of one another for the good of the institution as is the case in systems thinking.

When leaders make statements like this one that address "us" alone, it means they are set to guard their resources and knowledge to enhance their image. Since the linear-thinking mentality is ingrained in these leaders, they fail to appreciate that, no matter how well they perform, as a department no shine will come to them in any significant way if the rest of the departments perform dismally. They fail to acknowledge their role on the chessboard, where, unless the game is won, their level of participation will have been useless.

- There is no need for conflict

Can any critical thinking, which is very much part of systems thinking, take place without different views being criticized and sometimes dismissed? What is brainstorming, which is necessary when thinking in systems, without arguments? Arguments in systems thinking are expected to be constructive, but they are arguments all the same. Ideas contributed and even people will conflict as they try to argue their respective positions.

Nevertheless, all this is healthy and welcome, in fact, encouraged, if people within an organization are to spend resources addressing the right problems as opposed to symptoms, and if they are to find the most sustainable solutions to problems. American leadership consultant Edwin Friedman termed people who avoid discussing issues as they are in reality in order to avoid conflict, "peace-mongers," and said it was unfortunate they made it difficult, if not impossible, for people to address real issues for the betterment of everyone.

- You should do that like this and you will like it

First of all, you dictate to me how I'm going to do things, and then you force me to feign liking it? When people make a statement like this, they display their authoritarian nature, which is prevalent among leaders with the linear-thinking mentality.

On the contrary, leaders who think in systems have a collaborative mentality and will be open to hearing your views on how to solve the problem at hand. They are not domineering, because that would inhibit creativity and innovation, and it would also undermine the systems-thinking spirit of solving problems collectively.

While some people find it somewhat confusing, and sometimes overwhelming, when they are first introduced to systems thinking, others find it easy to embrace it as they have always used the systems-thinking approach to issues. Some have done so without consciously thinking about it, because that is how the families they were born in operated—collaborating and consulting.

Managers and other employees who find systems thinking overwhelming when it is introduced for the first time in their organization need not panic. The discomfort and uncertainty about how workable the new approach is happens to be normal, and those who become patient and accept the change with an open mind end up enjoying the sharing of responsibility that comes with the approach. Ultimately, when the organization realizes a drastic drop in expenses and reaps greater profits, courtesy of a smooth-running system, everyone is happy.

Best Problem-Solving Steps in Systems Thinking

If you pay attention to what happens in everyday life, whether it is in social circles, at home or even in the business arena, you will realize that many people, while attempting to solve problems, direct their focus toward what they think the problem is as opposed to looking at the entire ecosystem and trying to establish what could be disrupting the peace or the smooth-running of affairs. Fortunately, those who learn how effective thinking in systems is, and how sustainable the efficiency established in this manner is, want to continue polishing their systems-thinking skills.

Complex issues of a global nature have particularly encouraged people to embrace systems thinking, because, no matter how much people might try to address them through a linear approach, they cannot be solved. Problems such as those of global warming, for example, and environmental pollution, have served as a wake-up call, and now people generally acknowledge that every sector of the society and every community needs to participate in solving the problems involved. Today, those corporations that are doing marvelously well at a global level are among those who have embraced systems thinking and have kept on investing resources to ensure their staff understands the importance of having cohesiveness within the organization.

It is from this standpoint that communication has immensely improved in top performing organizations, as it enhances relationships between the various sections of the organization. In these organizations, every department understands how its everyday activities impact the customer, even a department whose staff members do not come face to face with the customers. As such, they do not need to be pressured to support the marketing and sales departments or to promote the company brand. All this is possible only when everyone in the organization has internalized the reality that they and their departments are systems with a crucial role, of ensuring the bigger system they are a part of succeeds.

If you, as an employee, are a system in a department that is also a system, and you join other employees and departments to form a bigger system, you are looking at a complex system in which you are a part of. The problems of this complex system are also complex because of the nature of where they emanate from, and so the most logical thing to do is to use a problem-solving approach that is designed to handle complex problems, and that is thinking in systems. The people who do not realize that, and stick to the traditional or linear problem-solving approach, unfortunately, end up bandaging problem areas and not really solving the problems. So they end up being known for a firefighting aptitude that is of little use to productivity.

It is very important that people learn how to go about solving complex problems, otherwise their attempts at problem-solving might end up introducing more problems into the mix. Albert Einstein said that there is no way you can solve a problem using the same thinking level that brought about those problems. It is, therefore, understandable that systems thinking calls for a broader approach to problem-solving, where you view the problematic area as part of a negatively-impacted ecosystem, rather than the part solely affected by the problem, and even solely responsible for it.

Although systems thinking has been embraced by big corporations and has become a talking point among many others aspiring to compete at a global level, the person credited with its introduction to the public is Jay Forrester, who worked closely with others at the Massachusetts Institute of Technology (MIT). Forrester described the idea of systems thinking as a discipline that helps you see wholes; a framework through which you see interrelationships as opposed to unrelated items. He explained how it enables people to see change patterns as opposed to snapshots of different scenarios.

This original concept of Forrester's, that explores life with its multiple systems interacting with one another and affecting one another, is as alive today as it was at its inception. Sometimes you may see a problem within your organization, but in reality it is being caused or exacerbated by external factors. As such, systems thinking can be applied to solve a wide spectrum of problems.

Before we highlight the best steps to follow when solving problems through systems thinking, it is important to understand the principles upon which this problem-solving approach is based.

Main Principles of Systems Thinking

The concepts explained here show the principles upon which systems thinking is used. They clarify the relationship that exists between a presumed problem and other system factors. They also help to give guidance on how best to observe the various relationships, with a view to finding an effective solution.

(1) Every system comprises parts that are interconnected. Therefore, any changes made to one of the parts, inevitably, affect other parts and the system as a whole.

(2) The system structure is the one responsible for its own behavior, and that means the system is dependent upon the connection between the various parts that form it, rather than on the individual parts singularly or collectively.

(3) System behavior is not easy to predict as it is an emergent phenomenon that keeps changing, and also because it is nonlinear. Remember, a nonlinear approach is where you can tie a particular cause to its direct effect. In the case of system behavior, it is not easy to predict just by observing the elements within the system or even the structure itself.

(4) A system's main dynamic behavior is controlled by various feedback loops, which are simply connections that cause one part's output to influence the same part's subsequent input. One thing that makes systems even more complex is the fact that a system has more feedback loops than even the number of its constituent parts.

(5) Wherever there are complex systems, there is counterintuitive behavior, and the emerging problems cannot be easily solved by conventional means. What is required are analytical methods and appropriate tools. The best way to deal with complex problems is actually systems thinking, because it matches the complexity of the problems as well as the dynamics involved.

After learning the above principles on which systems thinking is based, it is easy to appreciate the best sequence in which to implement.

- Meet with stakeholders

The reason it is important to meet the various stakeholders is to understand what vision they have about the situation. What would they like to see at the end of the day? When you all consider what the situation is like at the moment and what it should look like, you can then proceed to find out what the problem is, in which case you will be required to analyze the entire system and not its constituent parts separately. One can always use concept maps to demonstrate this. These are graphical tools that clearly show the organizational structure or knowledge structure. Such maps are great at presenting the elements, cross links, concept links and proposition statements within the system, and they also have room for examples.

- Design graphs showing behavior over time: BOT graphs

It is important to view how the problem has developed over a period of time, but people often err by looking at the problem as it appears at a particular time. Once you use the BOT graphs to understand how the problem has developed, you are likely to come up with a long-lasting solution, as opposed to when you try to fix a problem whose genesis and progression you do not understand.

The graphs representing the problem behavior over a period of time have a curve to represent that behavior change. The problem behavior is represented by the Y-axis while the time is represented by the X-axis. Using the graph, you can easily tell if the solution you are planning to implement has a chance of working and how well it is likely to do.

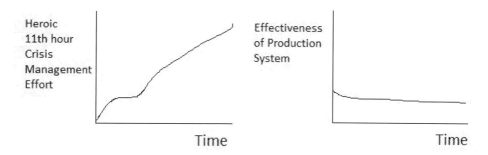

In the above BOT chart, the cost of management can be seen to increase over time while the efficiency in production can be seen to reduce.

- Design a vision statement

This vision statement should indicate what the target of the team is and the reason there is a problem in the organization. From such a statement, it is possible to learn what process needs to be undertaken in order to solve the existing problem.

- Describe the system structure

Once you have completed designing the team's vision and are clear on what needs to be done about the problem, it is time to describe the system structure with all its behavior patterns. The behavior patterns are great at helping to understand the problem better than before.

- Delve deeper into the problematic situation

After you have pinpointed the problem and established the system structure, you now need to look further into the problem with a view to understanding the complications within. To understand these complications, which will help you have a proper understanding of the problem, it is important that you clarify a few things, namely:

- What is the real purpose of this system you are trying to streamline?

- What mental method are you going to use?

- What is the larger system that this system is a part of?

- What is your personal role in this problem-solving undertaking?

Design an intervention

It is now time to use the information already gathered to initiate the phase of intervention. It is at this stage that you will try and find the connection between the problem and the various parts of the system, and then try to modify the problem so as to diminish its effect on the affected parts. Your aim here is to attain the desirable system behavior.

Farmers' Problems Solved via Systems Thinking

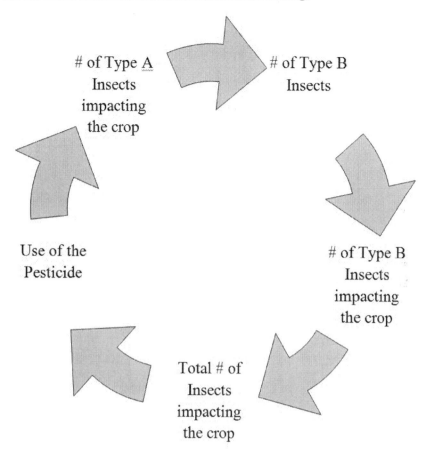

Expert Daniel Aronson's work is a good example of how systems thinking can be used to solve real-life problems. After noticing damage to crops and the accompanying costs to farmers, Aronson made the effort to identify the particular insects which were causing the damage.

People who use traditional thinking to find solutions to such problems would proceed immediately to use more or stronger pesticides on the farms with the aim of reducing the insect population damaging the crops. Incidentally, using that approach would reduce the insects on the farm and the crops would, indeed, not suffer more damage. Where this method fails is sustainability.

Spraying of crops in order to reduce damage from pests is only a short-term solution, and the people who insist on using this approach have no choice but to spray the crops every season, sometimes more than once in one season. Sometimes the problem is actually exacerbated by the use of pesticides because the type of pest being decimated may be one that controls the population of another pest within the same habitat.

The repercussions in the long run, therefore, include a drastic increase in the species of pest that used to be preyed upon by the pest made extinct by the use of pesticides. Soon, the crop damage problem resurfaces and the farmer is back to square one.

We have already explained that systems exist within other systems, and whatever happens to one system affects the others. In the case of the ecosystem, it has numerous other systems, including that of the insect population with its numerous species and the crop population with all its varieties. The question that needs to be examined is: what is the best approach to reduce crop damage in the long-term without destabilizing the natural environment?

Without human intervention, the ecosystem has a place for insects, crops and people, and the problem-solving method that can bring about sustained satisfaction is that which leaves all three existing in harmony. That is where the solution designed by MIT, as well as the National Academy of Sciences, comes in: Integrated Pest Management.

This approach to crop damage solution works by having pest predators introduced in the area, which means the pest predators reduce the population of the pests to a level where they cannot cause any significant damage to farmers' crops, yet the species is not threatened with extinction.

It is important to reiterate that complex problems cannot be dealt with using the traditional approach that tends to be linear in nature, although the approach seems to do fine in simple everyday problems. Complex problems can only be effectively solved by way of systems thinking, the approach that considers the entire ecosystem when attempting to solve problems, rather than considering the part that seems adversely affected in isolation or just a few parts of the system. This approach has proved it can relied upon to solve problems of the natural environment that include pollution, as well as others just as complex, such as poverty.

Chapter 14: How Systems Thinking Can Solve Social Problems

Although many social inventions are seen as not being geared towards profit-making, Evan Marwell, who initiated the nonprofit Education Super Highway (ESH), decided to adopt the mindset of an entrepreneur when he chose to work towards enabling every public school in the US to access broadband internet.

That mentality has become embedded in the minds of Americans now, and people generally believe that today, with the help of information technology, only one innovative company is required, led by a visionary entrepreneur, and the world will be transformed. As has been noted over recent years, when creative entrepreneurs brilliantly come up with innovations that are widely accepted, people are convinced the sky is the limit when it comes to how everyday lives can be transformed. Companies such as Google, Uber, Amazon, Facebook and others are great examples of how easy it is to overhaul people's social lives by being innovative and visionary.

However, the approach used matters when it comes to penetrating society because there are several stakeholders involved, and that is what Marwell acknowledged. He realized that, in order to achieve his goal of getting everyone connected to the internet, or simply to bridge the gap between those connected and those without reliable connection, it was important to involve a wide variety of stakeholders, gather fresh data and also work towards having federal policy changes.

He also knew it was imperative that teams were put in place to spearhead and guide individual states to establish the necessary systems. As for his part, Marwell found it necessary to act as a systems entrepreneur, in order to make adjustments to the entire system that was responsible for constraining internet access.

Understandably, it has taken some time for the community involved in the social environment to learn the best way for change to be effected through systems change, and that has made work more complex for the entrepreneurs involved. The entrepreneurs are called upon to acquire and use a set of tools, as well as to design and use a framework capable of addressing the complexity that is inherent in the prevailing situation.

One main area that has proven to be very complex is integrating innovations within systems of school districts, corporate structures, health departments, welfare agencies, etc.

In short, working on social systems brought about the challenges inherent when trying to modify systems for the better. Many of the social challenges are systemic, and they require critical levers to be engaged if the systems changes are to be effected in a way that alleviates perennial problems. Luckily, in this social endeavor, there were many stakeholders willing to help, and some of them even funded the projects involved. Gladly, their work highlighted a few pertinent characteristics of a good systems adjustment approach.

Characteristics of a Good Social Systems Adjustment Approach

(1) Systems thinking

It is important that an individual, or even an organization, establish a fresh solution or a set of solutions to be used to address a pressing social challenge. While this point seems to be obvious, it is meant to underline the importance of using the systems-thinking approach when designing business plans and other basic materials and tools to be used in addressing the problem. Whatever organizational change theories are to be followed, the people involved need to think in systems.

This is because, at the end of the day, what will make the team effort succeed in alleviating the pertinent problem, and not just put a pause to its effectiveness, will be a solution embedded into the affected complex system.

With Marwell's project, it was good he had the ability to view the entire scope of the complex system he was interested in changing. As such he was able to pinpoint the important pressure points or bottlenecks. There were two things that dawned on him as he was mauling on the system changes. Remember, he was interested in having the work of ESH integrated into the larger education system.

(a) The public, through governors and state governments, needed to be involved

It was important to involve these public officers and institutions because they held fundamental keys to the required change. It was not just their input that was required, but they needed to be frequently consulted and involved.

(b) Fiber networks needed to be expanded

Marwell was seeking to improve a system that had very many school districts; no fewer than 14,000. In order to scale the solution so that the enormous number of people could benefit, it was necessary to expand the existing fiber networks. The integrated approach employed in seeking a solution to expanded internet connections guided the entire ESH strategy.

(2) Research and analysis

Over and above having technical understanding of a certain solution and how it needs to be applied to a given problem, it is important for systems entrepreneurs to appreciate how the particular system or various systems they are attempting to change work. They need to understand exactly the factors that influence the way a system works. What Marwell did was create an "influencer map," which was meant to give him a more vivid understanding of all the participants from various fields—the federal government, communities, and different industries—because it was important that he engage them as important partners.

Marwell also designed a national diagnostic website that was referred to as SchoolSpeedTest, which was meant to build a larger body of data that was related to the problem at hand: the problem of inadequate access to the internet. He did that with full support from the Federal Communications Commission (FCC). His was a multi-sector participation, with a host of other organizations, no fewer than 100, involved.

(3) Effective communication

It is important to maintain transparent, compelling, effective communication with collaborative stakeholders. The same kind of communication needs to be maintained internally as well, and external audiences should be included too. That is the only way any efforts towards system change will become successful.

Towards this end, Marwell was aware it was important for him to raise awareness of the challenge at hand, in order to succeed in finding a solution. Consequently, he made a point of launching a public awareness campaign that focused on the issue of broadband access. He assembled a group of 50 CEOs, with a mixture of Democrats and Republicans, and got them to write to the FCC. That communication added to other letters from various governors and mayors, as well as leaders in education technology.

(4) Policy change

It is true there is polarization in politics, including in the United States Congress and Senate, but that is no reason not to try and seek policy change whenever you deem it necessary for public good. Marwell, when thinking in systems, realized the need to involve lawmakers so that there could be a fitting policy to help make the necessary system changes. Since the problem of constrained internet access had an underlying systems problem, it was critical that the system change be addressed in the right way.

Marwell rightly identified change to the Telecommunications Act of 1996 as being fundamental to the success of his problem-solving attempt. The Act had been used before to connect internet to most schools, as well as libraries, but it had not enabled those institutions to keep abreast of changing technology. After leveraging his network, Marwell was able to begin preparing his case for securing meetings with officials of the FCC as well as the White House.

(5) Assessment and evaluation

Assessment and evaluation is geared towards understanding the situation properly. It is important to know in advance the magnitude of the problem, what particular areas of the system are affected and how seriously. That helps to tailor the solution to fit the magnitude of the problem in specific areas.

This type of evaluation is different from the usual research that seeks to establish what a problem is and who is involved. It is conducted so as to create a consistent flow of data on an ongoing basis, which can then be used to guide strategy while increasing accountability. On the aspect of measurement, ESH worked on a constant basis with various partners. One of the best examples of the output of that collaboration was the report showing the status of each school's internet connectivity. The team, led by Marwell, also succeeded in pressing the FCC to be more forthcoming with data relating to internet connectivity and pricing.

It is a fact that today, Marwell, together with his team at ESH, have managed to drive unprecedented progress by using the systems-thinking approach. They managed to raise massive amounts of money, tens of millions of dollars, and it came from a wide range of partners. The big contributors included the Bill & Melinda Gates Foundation as well as Startup: Education. The team has also helped to influence reforms within the FCC's relevant policy, and consequently, from 2013, more than double the school districts that had internet access before had 100 kbps connectivity by 2016. Before, the percentage of connectivity was 30 percent of the school districts, and further evaluation showed that the percentage had risen to 77 percent.

A lot of people were involved in making this project the success it became, and that was because Marwell dared to use the systems-thinking approach, which enabled him to analyze the situation affecting an entire system and to seek a solution from the viewpoint of a whole system. He could have looked at the problem from a narrower perspective and reaped benefits as an entrepreneur, or translated it as a mere technical challenge, but he did not. Instead, he engaged critical thinking and sought to have the problem sorted out in a way that would root out underlying problems and all those emanating systems that appeared to be external. Thus, thinking in systems is the best approach to use for anyone interested in effecting genuine societal change.

Family Systems Therapy

Family systems therapy is a form of psychotherapy that employs systems thinking. It views the family as an emotional unit that is made up of interconnected and independent individuals. In systems thinking, different parts of a system are evaluated in relation to the whole unit. When this theory is applied to families, it implies that behavior is heavily influenced by an individual's family of origin. Families experiencing grievances within their family unit can benefit greatly from family systems therapy.

The Development of Family Systems Therapy

For years, Murray Bowen, an American psychiatrist, who was also a professor at Georgetown University, researched family patterns of people under treatment for schizophrenia, a mental disorder, as well as patterns of his own family of origin. He then introduced the family systems theory in the late 1960s. His theory suggests that individuals cannot be understood in isolation from one another.

Bowen, just like other psychologists of his time, was interested in designing subjective and objective treatment processes that would substitute customary diagnostic frameworks. He believed that all therapists had undergone some form of challenges in their family of origin, and this was reflected in their practice. Creating awareness of this could help therapists handle people in treatment better.

Traditional individual therapy addresses an individual's mental state so as to bring about change in relationships as well as other aspects of life. This approach only deals with one aspect of a system without realizing that each unit, in this case, each individual in a family unit, is interconnected. Bowen's theory suggests that it is important to address the entire structure and behavior of the larger relationship system as opposed to looking into one aspect. He believed that formation of an individual's character was influenced by the family and that changes in the behavior of one family member can influence how the family as a whole functions over time.

Family Systems Therapy Approaches

Generally, family systems approaches fall under three categories, outlined below. It is important to note that most forms of family therapy are based on the family system theory.

1) Structural Family Therapy

   This approach was designed by Salvador Minuchin. A therapist is able to examine the relationships, behaviors and patterns of a family as they are projected during the therapy sessions, and he can then evaluate the family structure using that knowledge. Therapists can further examine subsystems in the family structure, such as parental subsystems, by engaging the subjects in activities such as role playing within the sessions.

2) Strategic Family Therapy

   Strategic therapy was developed by Milton Erickson, Jay Halley and Cloe Madanes among others. Unlike structural family therapy, strategic family therapy evaluates the family behavior outside the therapy sessions. Strategist family therapists believe that rapid change is possible without going through an intensive analysis of the problem source.

   A therapist is able to examine family functions and processes, such as conflict-resolution patterns and communications, in a family's natural setting. Therapeutic techniques used in this therapy may include, but are not limited to, using paradoxical interventions or redefining and reframing problem scenarios. For example, a therapist can ask a family to take actions that are seemingly contradictory to their therapeutic goals so as to create the required change.

3) Intergenerational Family Therapy

   Murray Bowen designed this approach for treating not only families, but also individuals and couples. Consequently, this form of therapy looks at the family with a wider scope. It acknowledges that generational influences play a big role in an individual's behavior as well as the behavior of the whole family. Identifying generational behavioral patterns can help a family understand how their present challenges might be rooted in previous generations.

   Bowen used techniques such as discussing similar scenarios that a different family might be going through to show that the challenges are normal and are also experienced elsewhere. He also encouraged family members to take responsibility, rather than accuse or blame others, by responding with "I" statements.

4) Family Systems Therapy Genogram

   The family systems therapy genogram was developed by Bowen, and since then it has been adopted by many family therapists across the world. A genogram is a visual representation of a family's medical history and interpersonal relationships dating back at least three generations. It is a useful tool in highlighting hereditary traits, psychological factors, health issues and other significant issues that influence a family's psychological well-being.

Bowen used genograms not just for assessments but also to determine the ideal treatment for his patients. He would first interview each family member dating back at least three generations, so as to create a comprehensive family history. Equipped with this data, he would then highlight any prominent mental health and behavioral concerns as well as other important findings that were repetitive across the three generations. Presumably, the reason he collected data dating back three generations was that he believed that it took three generations for signs of schizophrenia to manifest in a family. However, as he furthered his study, he revised this estimate to at least 10 generations.

Concepts of Family Systems Theory

Bowens' approach is founded on eight main theoretical concepts. For you to understand the interconnection between these concepts, you need to have a thorough understanding of each.

1. A triangle

An emotional triangle represents the most basic molecule of a human relationship system. A team of two may be in harmony for a while, but it will destabilize once anxiety sets in. A third party introduced into the system reduces anxiety and relieves the system of some stress. Where three people can no longer contain the anxiety, larger relationship systems containing more people are established, forming a series of interlocking triangles that reduce anxiety and promote stability.

However, many triangles can also lead to conflict because each individual makes his or her own rules. Oftentimes, children find themselves in the midst of the triangle within their parents' relationships.

2. Differentiation of self

This concept describes the manner in which people respond and cope with life's demands. It refers to the manner in which an individual is able to separate feelings, thoughts and actions while pursuing personal goals. A person with a high level of self-differentiation is able to maintain individuality while still maintaining emotional contact with the group.

Highly differentiated people have a higher likelihood of finding contentment through their own personal progress while those with lower differentiation seek validation from people.

3. The multigenerational transmission process

This concept explains patterns of emotional process through multiple generations. One way that family patterns are transmitted across generations is through relationship triangles. According to Bowen, picking out a partner with the same level of self-differentiation potentially leads to certain conditions and behaviors being passed down to the next generations. If a couple with low self-differentiation has children, the children are likely to have an even lower self-differentiation. This can go on to the next generation and the next. To break this pattern, individuals should increase their levels of differentiation and prevent symptoms from recurring in other family members.

4. Family projection process

This is where parents, especially in a nuclear family, transmit their relationship difficulties and emotional anxieties on a child. The child then develops emotional issues and other related problems. Parents attempt to change the child, and when that does not bear fruit, they bring in experts to try and "fix" the child. Reports from experienced consultants who practice the Bowen family systems therapy show that parents who instead resolve their own relationship issues and manage their own anxieties see drastic improvements in their children.

5. An emotional cutoff

This describes a situation whereby a member of the family decides to emotionally distance themselves from the rest of the members. This may reduce tension and anxiety for a while, but it often leads to more harm than good. An individual may then seek other relationships to substitute for the estranged ones, but they may later project the same anxieties onto new relationships because of the unresolved issues they harbor.

6. Societal emotional process

Incidentally, emotional stress is not confined to individuals or families alone. Communities sometimes become collectively stressed. For example, society becomes anxious and unstable during periods of regression. Stressing factors such as scarcity of natural resources, overpopulation, economic forces, epidemics and lack of relevant survival skills can lead to regression in a society. This was the last concept that Bowen developed in societal emotional process.

7. Sibling position

Bowen adopted this concept from the research done by Walter Toman. It describes how siblings—the eldest, youngest and middle siblings—take on different roles in a family unit due to differences in expectations. Older siblings, for example, might be expected to act as young adults and take care of their younger siblings. These roles may be influenced by the sibling's position in the family.

8. Nuclear family emotional process

Bowen believed that the nuclear family experiences problems in four main areas: intimate partner conflict, emotional distance, problematic behaviors or concerns in one partner and impaired functionality in children. Anxiety often leads to arguments, fights, distancing behavior and criticism.

Benefits of Family Systems Therapy

Family systems therapy has proven to be effective for family, couples and individuals, and society is better for it. Family systems therapy has been used to treat many mental and behavioral health issues, such as schizophrenia, bipolar disorder, alcohol and substance dependency, personality issues, anxiety, eating and food issues, as well as depression. In general, the therapy has the capacity to deal effectively with those issues that relate to the family of origin.

Limitations and Concerns

Though family systems therapy is a widely used mode of treatment that is commended by both therapists and individuals in treatment, it has its shortcomings. One of the main challenges of the therapy is that it has a limited base of empirical evidence. Although relevant evidence is gradually growing, there is a need for more data to support the therapy's capacity, especially from objective sources.

Another weakness of the approach is the attitude of its practitioners towards their patients. Some mental health experts are worried that the seemingly neutral and unaffected response the practitioners of family systems accord their patients may encourage the harmful behaviors exhibited by the individuals, hence becoming counterproductive.

Why Systems Thinking is Best for Complex Problems

Systems thinking is a concept that views the world as one large system which is formed by numerous smaller systems. Systems thinking helps us to understand the nature of problems facing mankind and seeks to offer solutions.

In some situations, problems persist despite trying various methods of solving them. Sometimes, they not only persist but the situation continues to get worse. In such cases, application of systems thinking comes in handy to offer more sustainable solutions to the problems.

Systems thinking as a management approach looks at the interrelationship between different things and seeks to enhance that relationship for the sake of the bigger system. It looks at the patterns in various situations and how they are interrelated in the functioning of the whole system.

When Goals Need to Be Realigned

Sometimes we have a system that we think is failing but actually the system might be offering a solution to another problem, only it is not easy to tell at a glance. Let us consider a country's education system, for example. One might conclude that the education system is faulty because it does not produce a high number of people that society considers to be intelligent. Suppose that's an actual fact, meaning the education system creates a high number of people of only average intelligence. If we wish to change the products of our education system, then the best thing will be to redesign the system so that it produces a high number of intelligent people. The system that is deemed to be failing was probably not designed to produce the kind of people the society expects.

How to Deal with Feedback Delays

We often have difficulties solving problems because, in most cases, problems do not come singly. For example, in a city where there is an influx of gangs, the most common kneejerk reaction will be to increase police patrols to ensure safety. However, it might be that young people join these gangs after having unemployment challenges or abusing drugs. Systems thinking may help us see that, whereas the increase of gangs might look like an isolated problem, it is actually interrelated with other problems, some of them of a social nature and others of an economic nature.

In the above example, the increase in unemployment levels leads to frustration and increased stress. This forces youth to look for ways through which they can air their frustrations, find an expression of identity and rebel. This is the process that leads to the formation of gangs. We realize that gang violence and unemployment affect each other through feedback cycles. In such a case, the increase in one factor leading to the increase of another factor is called the positive feedback cycle. The opposite of this is the negative feedback cycle, in which the increase of one factor leads to the decrease in another factor. Using systems thinking, we can see the positive and negative feedback cycles affecting various things.

Feedback cycles are not always instant, and they may involve delays. In the case of unemployment, the result may not be immediate, but after a certain period of time when frustrations have reached high levels, it is when youth often decide to engage in gang activities. Just as in instances of medical treatment, symptoms do not end immediately after a person takes medication, but rather disappear gradually over a period of time. Understanding this is important so that one does not overreact during the period when we have delayed feedback.

Trying to solve complex issues without engaging the systems-thinking concept may backfire. A certain solution may look good, but because of feedback delays, we may not get the preferred results. Thus, we may end up making the situation even worse. Take, for instance, a patient who takes medication for a particular ailment but does not wait for the treatment to kick in, and so he ends up taking more medicine. Most probably the person will suffer from an overdose, making the situation even worse.

This often occurs when the solution is geared towards one part of the problem and not the problems affecting the whole system. Systems thinking helps us to understand that the unintended consequences may be a result of delayed or unexpected feedback. Patience is therefore required as long as you have followed the right procedures within the systems-thinking approach. Without reasonable patience, people may panic and begin trying to ape solutions used by other organizations, which might make your situation worse.

If we consider the various situations used as examples— youth gangs, the education system and a patient on medication—they represent a certain pattern of feedback cycle delays. Most other global problems also tend to have similar patterns in feedback cycles and delays. These similar patterns are what we refer to as systems archetypes. Systems thinking is crucial in helping us to view similarities between various problems by grouping them as common systems archetypes.

In every systems archetype, there are particular points that are crucial to effecting change. When we take situations as representing a specific archetype, then we choose the best systems-thinking approach to solve that problem by effecting change at the leverage point. The leverage point is the point of influence on a model where a minor change can cause a major impact to the whole system. Systems thinking can help us find the leverage points in various situations, and this may be useful in helping us solve various problems.

When Some Solutions are Counterintuitive

In most cases, the leverage point is not where it is expected to be, and sometimes it is counterintuitive. Take, for instance, a child who has low self-esteem and this leads to his poor performance in school. It may appear sensible if that child is pressured to work harder so as to get better grades. However, the intended results may not be achieved if the child falls into depression and self-hate, as that will make his grades sink even lower.

The workable solution is one that may actually be counterintuitive— that of rewarding the poorly performing child. In rewarding and encouraging the child, the child may end up accepting himself more and, hence, he or she will work in a more relaxed environment, which may ultimately lead to better results. Systems thinking may help us understand why some situations that may seem to worsen the situation might actually be points of influence in solving the underlying problem.

The Need to Put Individuals into Context

People's endeavors to understand how organizational events work and to intervene in the related processes are often influenced by certain biases. Often, people interpret happenings from their personal perspectives and within their own pre-existing mental frameworks. Essentially, therefore, personal or group needs, values, temperaments, motivations, personal styles and stages of developments of an individual or more people determine how situations are understood. It is not surprising, therefore, that the interventions introduced also happen to be of a personal nature.

Often, when leaders diagnose a problem, and of course, they do it through a personal lens, their quick solution is to demote someone or replace him, recommend coaching or therapy or something just as personal. However, Barry Oshry, a social systems expert, says there is a different lens through which individuals can view situations for their own benefit and those of others, and that is by trying to understand the context within which people exist and do what they do. Oshry refers to this view as the person-in-context lens, where any phenomenon is understood within the interactions of people as individuals or as part of a group; the systemic contexts within which the individuals and others exist.

This is the context that helps people gain a deeper understanding of the various phenomena as well as other leadership strategies, but, unfortunately, people usually ignore it. When context is understood properly, people reduce their tendency to apportion blame when systems seem not to be working as well as expected.

When people do not understand context, they tend to misunderstand events and also expend their energy in a misplaced manner. Unfortunately, many people lack the leadership competency that would enable them to see, understand and even master the systemic contexts within which they exist, as well as other people. In this section of the book, you are going to learn the consequences of being blind to context, as well as the fruitful possibilities that exist if only you understand and translate phenomena within the right context.

Incidentally, there happen to be four systems contexts within which you can view people's actions, namely, the top, middle, bottom and customer's context.

The Four Major System Contexts

The fact that we are discussing four common systems contexts does not mean these are the only ones within which people perform their everyday functions. There are other contexts, but these four are the ones that are crucial to your appreciation of organizational interactions. At the same time, they happen not to be just hierarchical positions.

They are also conditions that people experience on a day-to-day basis when interacting in an organization setup and elsewhere. People move in and out of these contexts with every event that they experience, and when that is taken into account, one can safely say that everyone belongs to one of the four major system contexts—the top, middle, bottom or customer context.

(1) The system's bottom context

When do people belong to the bottom context? This is when they are most vulnerable. In normal cases, people become vulnerable when unexpected events happen that people were not prepared for or are not able to handle well.

Essentially, people become vulnerable or find themselves in the bottom context when decisions are made affecting them in a big or small way, and they are not part of those decisions but happen to be on the receiving end. Good examples of such instances include when companies wind up business operations and close down, thus laying off employees; retirement benefits are reduced; the government introduces regulations restricting one's operations; initiatives that have been established are suddenly abandoned and so on. The fact that such decisions are made without involving the people they are going to affect most makes those people vulnerable, thus placing them in the system's bottom peer group.

(2) The system's top context

People fall under the system's top context when they have been assigned a leading role or a position of responsibility within a system or subsystem. It does not matter whether the designation of responsibility is within a division of the organization or at the top of the entire system, within a task force or a project, in a family setup, a team or even a class—it is still a top context.

The system's top context comes with complexity and accountability, and the individuals in those positions have massive input to handle, complex, and often difficult, issues to deal with, and they come from within, and also from outside the system. When you find yourself in such a position, you should anticipate issues that fail to be sorted out elsewhere to somehow float your way, and you will be expected to sort them out. You will also be expected to make complex decisions and take responsibility for them. Such decisions may be associated with the system type or culture, or even direction.

(3) The system's middle context

People are considered to be in the system's middle context when they are in between two conflicting demands, needs or forces, with each trying to draw the person's attention. For this reason, the system's middle context is said to be "tearing."

Good examples of instances when people find themselves in the middle tearing context include when one is torn between the interests of an executive group and the board of directors, between the interests of a supplier and those of the manufacturer, between the interests of a spouse and those of a child, etc.

(4) The system's customer context

People are considered to be in the system's customer context when they face neglect. Sometimes a customer wants a product or service but it is not forthcoming, and when it becomes available the price is exorbitant. Other times the price is not an issue but the quality of the product is wanting. There are even times when the product or service a customer wants is available and the price and quality are all good, but delivery is too slow for comfort.

In the context of systems thinking, a person does not need to be a customer for goods or services to be in the customer context. As long as there is something you want and you are at the mercy of other people, departments, organizations and so on, you can consider yourself to be in the system's customer context.

One important point that needs to be clarified is that, whatever the context you and other people are in, the movement in and out of those contexts is constant. So you should not view any one context as being permanent. The same person who has been identified as being within the system's top context can be found in the customer context during another system event.

Understanding People in Context

Often when people interact they do not reflexively think of one another as belonging to respective system contexts. Rather, they see themselves interacting as individuals. Even when it comes to someone requiring attention from another one, the person whose attention is required hardly considers the system context the other person might be in. In short, people are often blind to the contexts of others just as they are sometimes blind to their own contexts.

However, it is important that people make a conscious decision to consider other people's contexts when interacting with them, so that they can choose their actions with sensitivity. When people fail to appreciate the contexts others are in, serious misunderstandings often arise and sometimes people take inappropriate actions. Consequently, the system, whether it is at work or at home, experiences dysfunctional consequences.

At this point, it is important to understand the contextual principles that apply on a personal level.

Principles that Guide People's Actions

1) When people are blind to other's systems contexts

It is not possible to empathize with a person's position if you do not appreciate their context. As such, when people do not make the effort to understand the systems context other people they are interacting with are in, there is a high likelihood of there being misunderstandings and misconstruing of the other's actions. They even attribute wrong motives to them or respond in ways that put relationships in jeopardy.

Usually when people are blind to the context of others, the way they behave ends up adversely affecting everyone's personal effectiveness, and overall, the effectiveness of the system is adversely affected.

How Misunderstanding of Context Affects the System

(a) Tops and their arrogance

People in an organization can be disillusioned when there are people within the system's top context who behave in a discouraging way, essentially being blind to the context of others working under them.

Take the example of an employee who comes up with a brilliant idea and presents it to the management. You and everyone else in your group thinks the idea is great and has the potential to take the organization to another level. However, it happens that the people at the top do not even acknowledge the employee's efforts after accepting the documents. To them, the idea is just complicating an already complex world. In the meantime, the employee is looking forward to the day someone from the top communicates what they think of the idea. In any case, in your view, it is a brilliant idea that the organization is lucky to have received from someone.

Disappointment begins to set in when weeks and months go by with no communication from the top. As a result, you and the employee conclude the system tops are just arrogant people who have no courtesy when it comes to acknowledging good work, and that makes you both fume with anger. Worse still, from then on you become withdrawn and lose all the enthusiasm you had for being innovative for the welfare of the organization.

This behavior is not positive for the organization, and certainly your contributions (and the employee's) are likely to be lackluster from here on out. Yet this scenario played out in this unfortunate manner because the top's context was not understood. The top in this case was simply overwhelmed, not arrogant, and if that fact had been known, very likely you and the employee would not have developed a negative attitude that is now adversely affecting the organization.

(b) Bottoms with their resistance

It is not only the tops who are misunderstood. Bottoms are sometimes misunderstood, too, just because people fail to understand their context within the system.

Take the example of management coming up with a new initiative, which, according to management, should be exciting to everyone in the organization. In fact, it is an initiative that is geared to benefit all employees across the board.

Unfortunately, the initiative is given a cold reception by the employees within the bottom context, and management just cannot see how anyone within the organization can be indifferent to such a great initiative. In their disappointment on witnessing the employees' lack of enthusiasm, their conclusion is that their workers don't care and there is no need to bother trying to change anything, even if it's for their own good. The question is: is it true the employees at the bottom context are too far gone to be excited?

That's not necessarily true. It is only that they are in a vulnerable position, and so they respond in this manner in a bid to protect themselves from what they see as imminent disappointment. If there was an initiative they had been looking forward to and it never materialized, the failure served to fuel the group's vulnerability, hence their lack of enthusiasm toward the proposed new initiative. Had management understood the bottoms' position, they probably would have made more effort to allay their fears.

(c) The middle "weak"

When people find themselves in the middle context, they usually come across as weak and that is because the environment they are in does not allow them to be decisive. They owe allegiance to two powers that are not relatively independent, and the middle tries to portray a good image to each of them.

A good example is when you ask a middle to help out with something that is work related, which would take around half an hour. To you that is not something to think about, considering half an hour is even shorter than an employees' lunchbreak, and the season is not too busy at the company. However, instead of receiving a response such as what the best time is, the middle tells you, "Let's see … Let me get back to you."

That cannot fail to shock you, as the assistance you are asking is so commonplace anyone can do it, but it is only that the middle you asked happened to be at a workstation near your office. So you ask yourself, really? This guy needs time to ponder something so small and inconsequential? How weak can he be!

The reality is that you do not appreciate the person's context, and so you end up misjudging him as weak. He fears that being found in your office when you are not his direct supervisor will appear as if he is undermining his own supervisor. Yet you are also senior to him, and usually in charge of signing a certain category of requisition vouchers for him, so he does not want to outright turn you down.

(d) The "nasty" customer

Suppose you have messed up service for one of your customers by delaying delivery of his supplies. It is not the first time this has happened, and the customer is really furious.

The marketing department invites the customer to go out for lunch with two of its members, and in the meantime sends him a feedback form to complete. This invitation is a way of telling the customer the company knows it has messed up and wants to apologize. By taking time off to have lunch with the customer on the company's account, the marketing team expects the customer to understand that the company values him.

Unfortunately, the invitation just triggers another bout of complaining from the customer and he turns down the offer. To the customer, the company is simply negligent and probably sees his money like a drop in the sea, since the company has many other customers. However, to the marketing team and the rest of the company who know about the situation, this customer is just being nasty. How can he not accept the olive branch extended to him? The company has gone out of its way to show him they are sorry and that the mistake was inadvertent.

These situations can be avoided by following the leadership strategy of taking into account other people's contexts. Instead of labeling people nasty, weak and so on, you need to do what you can to enable them to accomplish what you want, or what the system requires them to do. It becomes even easier if you make an effort to make it easier for them to perform as required.

For you to succeed in thinking in systems, you need to keep in mind that tops are not necessarily arrogant, middles are not necessarily weak, bottoms are not necessarily resistant and the customer is not necessarily nasty. They appear the way you see them because of the context they are in. They are ordinary people trying to cope with their various contexts of accountability, tearing, complexity and neglect, and they can contribute very well to the system if they are understood and supported.

The reason problems within the system structure persist is that people keep reaching out and following that with reactions just because they take their relationships to be simple person-to-person interactions. In their continued blindness to system context, they elevate the incidence of top complexity, middle tearing, bottom vulnerability and customer neglect, yet that was never the intention. In short, it is important to always consider other people's context if you expect relationships to thrive and things to run smoothly without conflict or ill feelings.

In this regard, it is important to hold back any reflex to react when people give you answers you do not like or behave in the way you did not anticipate. Instead, try to understand other people's contexts as you interact and empathize with them.

At the same time, maintain your focus on what you want accomplished and be strategic in the way you move forward. Even as you appreciate those people's contexts, think about what you can do to win them over and get them to do what is needed.

2) When people are blind to their own context

When people are blind to the context they are in, they can easily fall into dysfunctional scenarios as individuals or as part of a system, and their relationships can be messed up. Often, people who are unaware of their contexts respond reflexively, behaving as if they have no part to play in being where they are.

Chapter 15: Appreciating People's Contexts for Profitability

The systems-thinking approach, as you may appreciate by now, is not about one person, one company or one unit that appears like a single system. It is a problem-solving approach that is multi-pronged in that it acknowledges the context in which the individual, the company or other entity functions. No single system is an island, and even an island has an ecosystem to interact or contend with.

In organizations such as businesses or manufacturing companies, the working structure is laid out to accommodate the top, the middles, the bottom and the customer. Whenever you are at the top and you ignore the middle, the bottom and the customer, something goes wrong with the entire visible system that is the company. The same case applies when individuals at any of the other levels ignore those in other levels of the structure. Just as bad is the case where the customer is ignored or taken for granted.

Yet the focus should not only be in entertaining the ideas of others and considering the effects of your actions on their functions. You also need to be sensitive to your own context. When you fail in the first instance, so that you end up not caring what people in other organization strata are doing, you end up misunderstanding their actions. Likewise, when you do not consider the context you are in, you are likely to respond to what is happening without being fully aware of the reality. Very likely your response will be kneejerk, and you will respond without much of a choice.

There are other principles besides being aware of your context and considering the role of other systems in the entire system that come in handy in helping to guide people's participation in development.

How to Streamline Group Operations

Wherever you work in an organization, there are peer groups, often in the form of the executive, middle management, members of staff and other lower cadre that include additional support staff. As if it is not sufficient to have divisions in the form of cadres, individuals bring with them their personal biases and prejudices, which are mostly influenced by individuals' affinities and antipathies. For this reason, people have a tendency to explain problems or failures in a group context as being the result of someone, either you or someone else, having done something untoward. In short, when a mistake is noted where a group is involved, personal issues are introduced. People sometimes go to the extent of alleging they are likely to be incompatible as group members.

Is it surprising, then, that people tend to look for solutions of a personal nature? After all, they are trying to solve problems which, according to their diagnosis, are of a personal nature. As such, solutions often come in the form of firing employees, rotating people to different work stations or shifts, seeking divorce, recommending coaching and, in the case of marital problems, therapy. In short, the approach you use to diagnose the problem usually informs the approach you use to alleviate those problems where two or more people are involved.

In reality, it is unfortunate that in many instances when peer groups break down, the reason is not based on personal problems but on people's blindness to context. Hence, those problems, once noted, are only made personal by the way they are handled, otherwise they are originally not.

Is it bad, then, to have an organization in some form of divisions? No, it is not. The issues an organization deals with, whether social, business or political are complex, and divisions are helpful when they differentiate individuals' or groups' different roles. In fact, without such kinds of categorizations, it would be quite difficult to handle the complexities of everyday operations and to accomplish responsibilities.

Danger of Being Blind to the Group's Context

People are sometimes blind to the context within which the peer group operates, and that makes the group often operate in a dysfunctional manner as opposed to working in harmony as one cohesive entity. When a peer group malfunctions in this manner, it causes the people involved personal stress, and the relationships among individual members of the group are weakened, and sometimes end up completely broken. In such circumstances, there is an unhealthy tendency to lose sight of the positive contribution other members are making to the group and the system as a whole. The dangers manifest in the form of:

1) Territorial tops

People within the top peer groups begin to view themselves as mere workers whose only responsibility is to accomplish the functions assigned to them. Yet, in reality, these group members are more than mere employees with a job to accomplish. They exist in a complex system, and the group that is part of that system is accountable for the success, or lack of it, within the system.

Whenever people are not aware of their contexts, they become vulnerable, not only to personalization of issues, but also to becoming part of a dysfunctional group. The whole dysfunctional territorial segregation begins with the individuals within those top teams choosing to handle their respective responsibilities their own way, without putting their role in the group context. Often, they will have adapted the inherent complexities as their own responsibility, seeing themselves as accountable for what transpires.

Much as differentiation is important, the rigidity that often develops from it causes problems, especially when those differentiations begin to be viewed as territories. In due course, individuals become masters of their own areas, gaining more and more knowledge about them and taking responsibility for what happens within those territories, while losing sight of what is happening in other territories of the same system. As people drive themselves towards narrowed specializations, they become less and less informed about what happens in other areas of the system and do not feel responsible any longer for the success or failure of those other areas.

That is how you find a workforce with a "mine" mentality, with people trying their utmost to protect and defend their respective territories. When things turn this way, the system is threatened and nobody can be certain about its future. To be able to work towards a strong system, it is important that the groups within a system work in liaison with one another, and the presumption is that the individual groups are, themselves, cohesive.

When people work towards the good of the system as a whole, as opposed to marking their territory and working solely towards the success of those territories, it becomes easy to discuss more pertinent issues such as the best culture for the system to develop, which direction to take in the future, what additional activities to introduce, the system's financial stability and prospects and so on. Nobody alone can answer such important questions for an organization, but answers can be found by thinking in systems and particularly by having individuals and groups thinking and operating within the right context.

On the contrary, when people take fixed positions within the organization, there is inevitable polarization, with silent conflicts emerging regarding which individuals or groups are more powerful than others. Some people begin to feel as if their contribution is not recognized, while arguments crop up as to who is responsible for system bottlenecks. Some groups accuse others of slowing down the system or holding it up, and confusing messages are passed down the system ladder. Redundancy of resources then becomes inevitable and potential synergies abort. In due course, tensions build intensively and extensively at the top level, and every conflict feels personal as individuals and groups try to exercise their controlling powers.

2) Fragmented middles

The peer groups in the middle of the system structure sometimes view one another as merely people. As such, they attribute their sentiments towards one another as mere indications of their varying personalities, values, temperaments, motives and the like. However, this is not all there is to it.

The middle category of groups is designed to exist in a tearing context, where each individual tears away from the rest of the peer group and tends to lead toward the people he or she is meant to lead, manage, service, supervise or even coach. The tendency to disperse is, therefore, simply an adaptive response to the preexisting tearing context. Individuals end up developing a mentality of "I" as opposed to "we," and at the end of the day the aloofness from one another dominates. Consequently, stiff competition develops among people within the middle groups, with each person gauging the others against themselves at a blatantly superficial level.

They assess each other's manner of speaking, manner of dressing, emotion, gender, skin color and so on. This tendency of fractionation isolates them and leaves them unsupported, and they often lack a peer group. They are among those people often caught by surprise when things happen, with a tendency to feel undercut by others like them in the middle category. This entire scenario is full of uncoordinated activity, and it jeopardizes any chance of potential enhanced synergies among the middles. It also makes it very difficult for the middles to have any collective influence in the organization.

3) Conforming bottoms

The peer groups at the bottom of the system often coalesce, and they have a sense of shared vulnerability. In fact, their tendency to coalesce is reflexive, emanating from their common vulnerability. As they keep together, they hone their "we" mentality, and that makes them feel less vulnerable, because, in that existence, their differences become insignificant, so what they end up feeling is a deep connection with one another.

Is that scenario all good? No, it is not. These people at the bottom not only coalesce, but they view other people outside of their peer group as a threat. Their uniting as a way to deal with their vulnerability makes them harden in that unique solidarity, and they view themselves as distinctly separate from everyone else. The issue of existence, therefore, becomes "we" versus "them."

Once that bottom cadre becomes wary of all other groups, it is not surprising to find them antagonistic. With conflicts emerging every now and then, pressure increases for every member of this group to maintain unity. Although the pressure is self-inflicted, anyone airing a dissenting voice is seen as threatening the "we" status, and those attempting to take up a different stance are pressured to discard their discordant thinking and toe the line. From the group's perspective, this is the only way it can maintain unity amongst itself. In fact, individuals are discouraged from holding differing positions on issues and instead are pushed to conform to the group's position.

The group pressure is so intense that the rights of individuals are infringed upon, including the freedom to choose and the right to exploit their potential as individuals. Obviously, with individuals' potential being under-realized, the system is bound to operate below par, which is an invisible cost to the system.

In addition, the continued resistance to change, even when initiatives are well meaning, is a great cost to the system. All those energies that are suppressed through group coalescence could be better utilized in developing the system operations. Everything involved in this kind of working results in an environment that is stressful for the group members, and their quality of relationships worsens. In the process, the group's overall contribution to the system keeps declining, and this, in turn, causes the overall system to perform dismally.

Does the bottom group have to operate in this manner? No, it does not. It is possible to transform a group with a vulnerable mentality to one that feels valued, thus making people take responsibility as both individuals and as a group. However, for this to happen, the people involved need to be aware of their context as individuals and also as a group, and then they can make the appropriate choice.

When a good leadership strategy is put in place, people are able to recognize the context of their peer group within the system. They are also willing to adapt to that particular context for the good of the system, as opposed to strengthening oneness to the extent of building isolation and territorial divisiveness that makes the group become dysfunctional.

It is also helpful when the peer group is developed into a robust system, with the capacity to make the individuals within the group strong while at the same time enhancing the relationship among the members themselves. As this happens, the group's contribution to the overall system improves and continues on a progressive path.

How to Strengthen Peer Groups

For peer groups to become and remain strong, it is important to:

- Understand the crucial systems underlying the existence of robust systems. This means it is important for you to know, acknowledge and understand the systems with unique capacities to exist and even develop within their own environments.

- Learn how context influences systems processes in a manner that limits the effectiveness or efficiency of the respective peer groups.

- Become a master of systems processes, because, in any case, any single peer group is capable of transforming into a robust system.

What a Robust System is

A robust system is one with the capacity to differentiate and homogenize, even as it individuates and integrates. These terms do not portray matters that are excessively complex, as will soon be explained.

A Differentiating System

As for differentiating, it means the system has the capacity to develop in form as well as in function, which enables it to interact with its environment in a more complex manner.

A Homogenizing System

A homogenizing system is one that facilitates information process sharing as well as capacity sharing within the entire system.

An Individuating System

A system is said to be individuating when it encourages individuals, as well as individual peer groups, to operate separately and to independently penetrate the environment for causes that are helpful to the self and the system, including experimenting and testing matters, as well as developing the potential of individuals and groups.

An Integrating System

An integrating system is one that enables individuals and system units to work together while sharing information and supporting one another. In the course of all the system parts coming together, they modulate one another's activities for the welfare of the entire system to which they all belong.

It is important to appreciate that, irrespective of whether people see their context or not, peer groups will automatically adapt as a matter of reflex. Unfortunately, sometimes the patterns formed after the reflexive adaptation end up reducing the effectiveness of the peer group in the system. This is because they rely heavily on particular processes as they ignore or suppress others. One of the greatest advantages of understanding context is that you develop the capacity to strengthen the respective peer groups by identifying the ignored processes and enhancing them, making them great contributors to the system functions.

How Groups Fall into Top Territoriality

Top groups fall into territoriality through differentiation and individuation that do not go hand in hand with homogenization and integration. When top groups are within the context of complexity as well as accountability, the quickest response, that also happens to be reflexive, is to differentiate and individuate. This means they develop various forms and processes geared towards helping them cope with the complex aspects of the system, as well as to enable each of its constituent parts to work independently. This behavior by the different parts of the group develops in a bid for each part to pursue its own strategy and approach. Having the different parts of the group develop this way is a positive thing.

However, a problem arises when these top groups do not embrace homogenization and integration, which would have had the effect of balancing the combined effects of differentiation and individuation. With this lopsidedness, the groups inevitably enter into destructive territoriality. It is, therefore, imperative that leaders identify means of developing a robust top peer group.

The best way to attain a good balance and develop a robust top peer group is to consciously work towards more homogenization and integration. It does not work well to do the converse—working towards reducing the tendency to differentiate. With more homogenization and integration, the group members are able to share high quality information with one another while investing time into sharing each other's experiences. They are also able to collaborate better on projects, as opposed to individuals working in specialized arenas, and learn from mutual coaches where all members within the top peer groups are inclined to help one another succeed.

It is this kind of homogenization and integration of group activities that enhances the group's capacity to perform well. As such, it is safe to conclude that the best way to improve the performance of top peer groups is to use homogenization and integration to strengthen differentiation and integration.

How Groups Fall into Middle Alienation

The best way for a peer group to fall into middle alienation is by way of individuation in the absence of integration. Middle groups within the tearing context have the tendency to individuate, and that is actually their reflexive response. They separate themselves to function independent of the other groups they are meant to serve, as they supervise, manage, coach and support them in various ways. This is translated to be an adaptive response to the already mentioned tearing context.

Whenever there is no integration to strengthen individuation, a pattern develops that is fractionated, which is perilous to the entire system. That then begs the question of how leaders can develop a middle peer group that is robust. The challenge that exists here is that of ensuring the peer group integrates more. It is not important to work on trying to make the group individuate less.

In a bid to integrate more, the group members ensure they meet on a regular basis as members of the middle peer group, and they share information that is well assembled from the system. They take the opportunity to use their shared intelligence in a way that helps them identify any system issues that exist, and also share the best system practices to deal with the problems. They work together to effect changes that are otherwise difficult for them to implement as individuals. In short, the best approach for a middle peer group to exercise its power is to use integration to strengthen individuation.

How Groups Fall into Bottom Conformity

The way peer groups find themselves within the bottom conformity is by way of homogenizing and integrating in the absence of individuation and differentiation. In the shared vulnerability context, the group's reflexive response is coalescence, this being the process by which the group maintains its unity through homogenization and integration. In homogenization, emphasis is put on the commonality of the group members, in the meantime suppressing differences among the members that have the potential to divide them.

The group members at the bottom support one another in their common cause, as they find themselves coalescing due to shared vulnerability. In short, this coalescence is the group's reflexive response to their vulnerability. Whereas this dynamic is good in enabling the group to work smoothly, it becomes a problem when there is no individuation and differentiation to balance the homogenization and integration, because the group begins to stifle divergent views and efforts, while cultivating destructive conformity. The question then arises of how best to develop a bottom peer group that is robust.

The challenge that exists for the bottom peer group leadership is how to go about strengthening the members of the group by getting them to embrace differentiation as well as individuation. In differentiation, they will allow multiple approaches to be tried out to deal with their vulnerability, while in individuation the members will be asked to venture outside their peer group to assess what unique contribution they are capable of making.

In other words, differentiation and individuation need not be viewed as threats to unity, because there is a way they can be used to strengthen the goals of the group rather than jeopardizing them. In this case, rather than putting emphasis on the strength that exists in unity, emphasis is placed on the strength that exists in diversity. Overall, the best approach to enhancing the bottom peer group processes is to strengthen homogenization and integration while individuating and differentiating.

How to Overcome the Perception of System Blindness

Ordinarily your consciousness is influenced by the way you relate with other people, and consciousness means how you experience or view them. For instance, you may suggest that a particular strategy you deem fitting be introduced to a group, and receive a response from the manager that indicates the strategy cannot work with the group. Then, when you ask why it would not work, the reasons might be associated with the group members' temperament, needs, motives, maturity or lack of it and so on.

If you critically consider those reasons, you will notice they are linked to the manager's personal opinions. So he has been considering work-related solutions through the lens of his eyes. That is how the top peer groups end up viewing others through the "mine" mentality, the middle peer groups through the "I" mentality and the bottom peer groups through the "we" mentality, all being solidly convinced that their views are reality. Any person trying to convince you that other people can be any different than you see them seems wrong, because there is no chance of you feeling differently about others as long you have your unique mentality.

You think any suggestions to the contrary are farfetched. Yet these experiences you feel are solid are actually illusions of your systemic blindness. If you're obliged to change your relationships, your feelings are also bound to change.

Tips for Total System Empowerment

The tips provided here are important to everyone, irrespective of their position within the system or the organization.

(1) View yourself as constantly moving in and out of the four system contexts of top, middle, bottom and customer.

(2) Be confident that you have the capacity to make a positive impact on the system and to strengthen it whatever your context position in the system, helping it optimize on available opportunities. Also know

that you have the capacity to help the system survive and develop, and to withstand the prevalent environmental dangers.

(3) Understand your roles depending on the context you are in.

a. When in the top condition, your system potential is to operate as developer.

b. When in the middle condition, your system potential is to function as integrator.

c. When in the bottom condition, your system potential is to function as fixer.

d. When in the customer condition, your system potential is to function as validator.

All this is possible if only you avoid responding reflexively, evading responsibility when in the top, trying to take up high responsibilities when actually in the bottom, missing a grip on connectivity when in the middle and assuming delivery systems are the only ones responsible for having deliveries made. When you make a point of enhancing system power within all the system conditions and as they relate to one another, the system becomes stronger and you can safely speak of having total system power.

Chapter 16: How to View Your Company as a System

It is safe to say whenever you look at something, whatever that thing is, it is a system. This is because there is more to what you are seeing on the surface to that thing. As in the example of a human being who has different biological and physical systems, each with unique characteristics, every other system has other systems that make it function, and others help it survive. The world, which is itself a system, comprises systems within systems, and the company is one such system that is within the economic system and operating within the ecosystem.

Different systems have their vulnerabilities; only the degree and nature of vulnerability varies. For example, banks can collapse and put other systems, including their respective countries, in positions of vulnerability; health insurance rates can hike sharply and suddenly and put companies, businesses and families in a vulnerable position and a lot more can happen and exacerbate system's vulnerability. When it comes to companies specifically, even the rate at which technology changes can make them vulnerable. If not put in check, these failure triggers can bring down systems and cause companies to begin regressing instead of progressing.

Unfortunately, when a company fails, the general conclusion is that management has dropped the ball. It is, therefore, imperative that executives work together with politicians towards regulatory reforms so that it becomes easy to build and strengthen social systems, including companies, and to keep them stable. Many people know companies as economic entities, but they may not recognize them as also social systems, and complex ones at that.

A company is also a living community, and you can appreciate this when you consider that its operations are carried out by people, and decisions about it are also made by people. It is not surprising, therefore, to learn that there have been management thinkers considering how to run company systems since the early years of the twentieth century. That notwithstanding, there are still leaders who have an inclination towards machines, and even when they are dealing with systems involving people their behavior is similar to that of pulling levers and pushing buttons. Such leaders often make unilateral decisions, with massive impacts on the company, without bothering to assess what other people's feelings are likely to be and how the changes that follow are bound to influence the response of the company as a whole.

Nevertheless, there is still a good proportion of management within the corporate world that recognizes the need to involve people in its decision-making process. This is particularly important because today's global business environment, which is already complex, is changing at a very fast rate, and for individual companies to operate competitively, there is a need to enhance relationships amongst people and among systems. The overall picture should actually be one where management has a systems-focused view of the business world. With the right outlook, efforts will be made to enhance system processes and designs, and to recognize and respect people's interrelationships.

Why People's Involvement in Decision Making is Important

Although there are different intellectual approaches to understanding systems, there are some cross-cutting issues among all the thinkers.

1)   Static solutions can be a source of more problems

This assumption takes into account the reality that companies are complex systems that keep changing, and therefore employing static solutions to system problems is not only bound to fail in alleviating the problem, but is likely to cause system destabilization. When someone dashes in to make a quick fix, the move often delays or locks out any further attempts by management to seek a long-lasting solution to the problem. The best recommendation, therefore, is for companies and other organizations to be dynamic in a manner that permits flexibility during unexpected developments.

2) Companies can benefit from trials

Anyone who has had opportunity to think and act creatively knows not every trial works. However, the best solutions are often found after persistent trials, even if those trials keep producing errors. It is those trial-and-error efforts that end up producing sustainable solutions to problems because the system vulnerabilities will have been learned and taken care of during the trial, such that, finally, the system is designed or modified to become self-correcting.

3) It is important to understand individual parts of a system

Whereas stress is placed on looking at an organization as one system, the only way that whole system can work well is if its constituent subsystems or parts are operating optimally. After all, any change on any part of a system ends up affecting the entire system through a ripple effect. Even changes to the environment of any of the system's parts, or even other subsystems associated with it, can also produce ripple effects.

It is noteworthy that all these assumptions appreciate people's participation in the making of decisions affecting the system. In fact, contrary to what people initially thought, which is that systems thinking is best for senior executives, engineers and others like them, it is actually very useful to marketers, designers, ordinary workers, etc. In short, thinking in systems is best for everyone, irrespective of their field or cadre. When people think in systems, they not only learn how the system works, but also sharpen their creativity and utilize it to improve and strengthen the system.

The company issues that should be considered when employing systems thinking are the same ones befitting small enterprises, because, whether the organizations has strong clear structures or vague ones, human interactions need to be nurtured. In fact, the same can be said of government institutions, because, although they do not involve profit-making, they do give value to the public in the form of services, and to do so they must consume resources.

What various experts have realized is that, whatever form of organizational structure exists, it is important to find a way of introducing systems thinking to the entity, so that any improvements made can be sustainable. One example that shows how positively and widely systems thinking has been accepted is the case of the US National Commission on Terrorist Attacks' report, which noted that imagination is not a talent ordinarily associated with bureaucratic bodies, hence the need to find means of embedding it within the bureaucratic processes and making it part of the institution's routine.

Important Lessons for Company Managers

Any managers enthusiastic to see their company succeed will be looking for ways to improve their companies' day-to-day performances at a conceptual level as well at the level of practice. In addition to soft management skills, they also need to be good at thinking in systems, especially of their capacity to strengthen relationships. They also need to go beyond enhancing employee-centered visions and embrace a systems-oriented vision. In any case, the workforce is a system that is part of the company, and the success of either has a positive impact on the other.

Companies seeking excellence usually emphasize constant learning, not only for individual employees but also for the entire organization. The presumption is that those people who are nearest any particular part of the system are the bearers of the most valuable and relevant expertise, and if they are not, then they require further learning. It is up to company CEOs to facilitate learning as well as easy adaptation and improvement, and the best way to do this is by creating an enabling environment. Such an environment is envisioned as being free of fear, with the capacity to provide training and appropriate tools to employees.

Once employees are knowledgeable and have the right skills, they are able to identify existing, as well as potential, system problems, and to think critically about how best to improve the system.

Company leaders are also expected establish routines that have the capacity to test currently held assumptions, and to try and anticipate potential needs. For them to achieve this, it is important that they pay enough attention to company processes. In any case, for appropriate changes to be made, whether in the form of physical system changes, administrative changes or innovative leaps, the leaders must, of necessity, persist in driving continuous improvement.

Professor Russel Ackoff, an expert in management science, has warned in his writings against organizational silos, or the mentality prevailing in some organizations where individual departments hold onto information and guard it against being shared with others. The same professor has also warned against organizational fragmentation, where service delivery is done from different points that are poorly coordinated, resulting in poor performance all around. The converse of that is what the professor recommended, which is, essentially, systems thinking.

When leaders and everyone else in the company thinks in systems, they feel free to share information as opposed to hoarding it within departments and using it selfishly. This shows there is appreciation in the company that, although the organization has different departments and systems, the success of one or more of them will not mean much unless the remaining departments and systems succeed, as they are also part of the main system. In short, there needs to be free information flow within the organization if the system is to succeed without bottlenecks.

Professor Ackoff himself, in his career, avoided confining himself to one narrow field of specialization, ending up studying architecture and philosophy and capping those achievements with systems sciences and management. He also did a lot of research and taught at the University of Pennsylvania. Ackoff advocated and practiced broadness of ideas and experience, and that is what you find in systems thinking where every subsystem is expected to be in the know about what other subsystems working alongside it are doing. This is important because, once everyone knows what everyone else's part is in the running of the company, it is easy to predict the impact every single action or process they undertake will have on other people, departments or systems.

Going by Ackoff's way of thinking, a company should look for sustainable solutions to problems and not rely on simple solutions that are repeated periodically. This is because, just as systems are interacting, so are the problems found within organizations. So, if simple solutions are sought and applied to the areas where problems manifest, other interacting areas will only be soothed, and soon, a problem related to the earlier one will emerge from another system point. Even as Ackoff recommends solving company problems globally, he also advocates creating systems capable of learning and adapting.

He said that, contrary to what many people assert, experience is not really the best teacher, and, in his view, it is not at all a good teacher. The reason for going against the grain and dismissing experience as being the best thing to rely on is that it happens to be very slow in its teaching, and it also lacks precision. Often, he said, the teaching experience provides comes with lots of ambiguity.

According to the professor, organizations should seek to learn as well as adapt by way of experimentation because it is faster, less ambiguous and even more precise than experience. In order to run seamlessly, a company needs to design systems that are run experimentally, instead of waiting to learn experientially. Ackoff then laid out what he considered the best method for implementing interactive planning. This method consists of an organization design that is idealized, showing how major stakeholders can redesign the system or rebuild it in case it is abruptly destroyed. The method is technologically feasible.

The idealized design created by Ackoff in the early years of the 1980s was later utilized by Clark Manufacturing Company at a period when the company's market share of its earthmoving equipment was being encroached upon by machinery of higher quality and lower prices from Japanese manufacturers. The company sought the services of Ackoff after acknowledging it did not have the necessary organizational capability to beat the stiff competition.

What transpired next was the acknowledgement that a system can work with others for its own good. In this case, the company realized it would be better off in terms of capacity and capability by joining forces with Euclid, a relatively small company dealing with trucks, as well as Volvo. Many managers were skeptical at the time that the system change would work, but the joint venture was initiated, and it proved to be a success. Effectively, the problem of unprecedented encroachment of the company market share was solved by developing a united management group comprising otherwise independent enterprises. This was proof that it is possible to work successfully through cross-cultural management.

At the end of the day, company management should not look to find a panacea for all its problems, but instead seek to establish a design process that is adaptive and always evolving. In Professor Ackoff's view, this is a design process with the capacity to manipulate sections of the company, with the goal being to improve the performance of the company as a whole.

How to Think and Interact Better

Another expert who took Professor Ackoff's approach seriously was Peter Michael Senge, an American systems scientist who is a trained aircraft engineer. He emphasizes the need to focus on blending technical elements with those of behavior within a company setup. He also emphasizes the need for individual company employees to embrace new ideas and perspectives and free their minds of any prejudices they may harbor both at conscious and subconscious levels.

According to Senge, however you see an organization working is a reflection of the thinking of the people involved with it, the way they work and also how they interact. In short, it is important to choose the kind of thinking that addresses problematic issues comprehensively and in a way that produces sustainable solutions. That is why systems thinking, within which situations are critically analyzed before solutions are sought, is recommended.

Senge happens to be the founder of the Society for Organizational Learning at MIT, where he has also served as lecturer, and he says as long as there is no trust among people working together, they cannot manage to build genuine aspirations and even mental models. Needless to say, if a team or a group is to produce the best results possible, they need to have common aspirations. Systems thinking in a company allows for this as information sharing about processes, problems and everything else is open and freely available to all involved. According to Senge, there are some disciplines people should embrace in order to achieve organizational learning.

Five Disciplines Best for Organizational Learning

(1) Systems thinking

Systems thinking is being thought of as a way of learning to appreciate those forces of work acceleration and equilibrium prevalent in complex systems, the manner they interact as time goes on and also how they can be used to gain leverage.

(2) Personal mastery

This is a variation on self-actualization, as advocated by Abraham Maslow. It taps into the massive creative potential people have by aligning their personal aspirations with the organizational goals, while keeping a clear view of the prevailing reality.

(3) Mental models

These involve the attitudes and cognitive habits, as well as beliefs, that people within a group have, which end up shaping the group's perceptions of the world and influencing how the group acts.

(4) Shared vision

This refers to the collective aspiration that has already been voiced, and which drives the collective endeavor to pursue worthwhile goals.

(5) Team learning

Team learning encompasses free dialogue that has no preconceptions, as well as different forms of group activity that are carried out candidly and in great depth.

How to Change the Blame Game to Accountability

When an error occurs in an organization or institution, in most cases it tends to look for someone to blame instead of looking for solutions. People tend to shift blame from one party to the other, as the people we blame often also target others for blame. These tendencies are very unfortunate because they discourage people from learning from those mistakes. Once people are blamed, the natural tendency is for them to go on the defensive. The truth is that blaming and the defensiveness are not good for the individuals concerned or for the organization. If the blame is between departments, the same applies. Neither of the departments benefit from the blame or defensiveness, and instead relationships are likely to become strained.

Such a state of affairs is not healthy in any organization because most people tend to focus their energy, which could have been used in more productive ways, looking for people to accuse of the mistakes which might have occurred. When people are blamed, they tend to fear asking questions and even making contributions, yet their contribution may have been valuable to the institution. The strain on relationships between members of the institution may escalate and persist, and this may lead to reduced productivity by the individuals involved and, of course, the whole institution at large.

Blame games are also a hindrance to effective problem-solving. It is therefore important that we endeavor to ensure that we desist from blaming one another, and instead seek to collectively find solutions.

One good way of reducing the incidence of the blame game is by promoting accountability. Accountability involves letting each member know what tasks they are supposed to undertake, which methods they should use to accomplish those tasks, what the time span is within which they are expected to accomplish the task and what standards they are expected to attain.

Differences Between Seeking Accountability and Blaming

- Accountability, on one hand, seeks to encourage people to learn from their mistakes, learn what might have caused the mistakes and seek out the best solutions to the problems so as to avoid repeating the same mistakes in future. Blame, on the other hand, seeks to embarrass those who are responsible for the errors that might have occurred. Blame does not provide a long-lasting solution to a problem but offers a temporary and simplistic solution to a complex problem.

- Accountability seeks to make people learn from the problems and to use that learning to improve on performance, increasing productivity. Blame does not help as it seeks to punish those who might have caused the errors.

- Accountability encourages openness, learning and innovativeness, while blame breeds mistrust and acts as a hindrance to learning and development.

- Accountability seeks to make people understand the problem while blame seeks to find the culprit—the person who caused the problem.

Consequences of Blame to An Institution

Blame games are often used to shift responsibility from one party to another, such as from one person to another, from one department to another, management to external stakeholders and so on. This approach, in most cases, fails an institution because most employees will view it as a punishment. In most cases, the employees will be tempted to hide information in order to avoid being blamed, and this tendency results in more process errors.

Trying to find people to blame is always the easiest way to evade dealing with a problem. In such cases, problems will never be solved but swept under the carpet until they resurface again and the cycle of blame game continues. Blame also causes fear among employees, and this curtails not only their personal growth but their innovative abilities.

Blame games offer a short-term solution to a complex problem. Blaming others only absolves the blamers from any wrongdoing, but it does not make them any better. For one, once they are exonerated from blame at the expense of others, they do not examine themselves to see how they may have contributed to the occurrence of the problem.

How to Change from Blame to Accountability

There are three levels that we can use to change from being blamers to seeking more accountability. The first level is the individual level where people must change their perspectives towards blame. People also need to encourage accountability amongst themselves as team members or members of a group. This enhances relationships at an interpersonal level, and people end up seeking answers from one another as well as solutions to problems.

Finally, institutions need to have a clear channel that can be used to solve disputes and to offer a forum through which problems can be aired. The same avenues can be used to discuss challenges encountered in the line of duty. Such channels are great in creating an environment that is conducive to work, and because of them, relationships are nurtured and problems solved harmoniously. At the end of the day, productivity levels rise, and everyone is happy.

1) Changing from blame to accountability as an individual

Such change can only come through a change of mindset. There are various ways to change your mindset from a blamer to a person who is accountable, and they include:

• Always remember that others are doing what they think is right and should be done. In short, according to the thinking of those people being blamed and the situation they find themselves in, they did what they deemed best.

• Ask yourself about your contribution to the behavior an individual exhibits. Probably the other party is behaving in a certain manner so as not to offend you.

• Put yourself in the other person's shoes and try to understand why they are behaving in a certain manner. Avoid being judgmental. Instead, practice seeking to understand the situation the other party is in.

• Be willing to be held accountable for roles that you undertook.

• Ensure that you understand the established goals and the most appropriate means to accomplishing them.

2) Changing from allocating blame to being accountable at the interpersonal level

Before tackling any task, it is important to know in detail what the task entails, the expected methodology to complete it, the set standards and the expected deadline.

It is also important to ensure that there is a clear plan between the employee and the employer on how to accomplish said task. This helps in that. when a problem occurs, the parties involved have a reference point from where to identify what went wrong.

It is also helpful to raise issues of concern as soon as they arise so that they can be addressed immediately. Piling up issues of concern only creates a ticking time bomb, which is bound to explode at the slightest provocation.

As you tackle the given task, it is important to check on the agreements from time to time, as well as the working plan. This way, you will be able to tell at every stage if you are producing the desired outcomes or not. Definitely, people who seek accountability are bound to express the challenges they are facing in the course of duty, which can then be solved amicably by accountable employers.

How to Conduct an Accountability Conversation

- When a problem occurs in an institution, seek to understand whether the people you are working with are interested in finding a solution to the problem.

- Create an environment where members can air their problems and commit to listening. Provide assurance that the conversations will seek to offer solutions to the challenges being faced and not apportion blame.

- Inform the participants of the intent of the conversations, and when it is time to discuss the problem, analyze the problem from a systems perspective. Let the members understand how their individual actions reinforce each other to the benefit or detriment of the institution.

3) Changing an institution from allotting blame to being accountable

In most cases, individuals in an institution do not channel their complaints appropriately but opt to share them with others. This offers a temporary relief to the problem but not a real solution. However, if the problem is channeled to the right personnel and is addressed appropriately, then a solution that will help avoid such a problem in the future is likely to be sought.

It is good to share the challenges that one faces in the course of duty with others, not with the intention of shifting blame to other people, but of seeking appropriate ways to air the grievances through the appropriate channels. Telling the most suitable person about the problem is likely to get it resolved quickly and effectively. That is what accountability seeks. As you share the problems encountered with a third party, avoid talking negatively about another person because this is likely to create hostility amongst individuals.

Seek an audience with the appropriate personnel and start by reaffirming your commitment to working together with them and offering to help find solutions to their problems. Be open-minded as you engage in the conversation and express the assumptions and feelings you might have on the situation.

Finally, it is important that you offer your suggestions on what you think is the best way to solve the problem. Remind the other person that discussing the problem is part of learning, and also the beginning of seeking a solution. Let them know that discussing the problem helps to create a healthy working environment, which in turn enhances productivity.

How to Establish Long-Term Solutions to Problems

Systems thinking tries to break the vicious cycle of blame, which, as we have noted, is a quick fix to a complex problem. In order to avoid situations that cultivate blame, it is good to have some guidelines, which individuals in an organization at all levels can follow, and which will condition people to always seek a solution whenever a problem arises.

Guidelines to Help Find Long-Lasting Solutions

• Seek to understand what causes employee ineffectiveness and low productivity.

• Seek to understand an employee's workload before giving them another task.

- Where more than one task is to be completed, let the employee know which takes priority.

- Provide adequate resources, inclusive of finances, to the employee, to enable him or her to accomplish the task up to par.

- Ask for feedback to ensure that any challenges are addressed in a timely manner.

- Offer clear communication. This includes requests, orders, commands, suggestions, compliments, complaints, instructions and inquiries. Whatever it is you want another person or other people to know, let your communication be as clear as possible.

Making a shift from blame to accountability is of fundamental importance to the efficient management of systems, primarily as it makes it easy to find long-lasting solutions to problems.

Chapter 17: How to Change Systems by Changing Mindsets

How your thinking process works in relation to the context you find yourself in can determine the kind of personality you develop and the values you hold dear, and consequently chart the way you act. This deduction is supported by findings in neuroscience that have shown that sensory experiences, as well as actions, alter a person's physical brain structure. In addition, other studies in neuroscience have also shown that thinking does the same thing, altering the brain's physical structure.

When you concentrate on things you should be grateful for, you are able to rewire your brain so that it develops an appreciative attitude. Likewise, if you focus on imagining yourself playing on the piano as a five-finger exercise, your imagination itself has the capacity to enlarge that section of the brain responsible for manipulating fingers.

Professor Elaine B. Johnson, who is Huron University College's honorary fellow and is renowned for her interpretation of brain research, says thoughts alone can alter your brain's physical structure, and that gives further credibility to the position already held by proponents of systems thinking and organizational learning.

There are strategies that can help people to have a productive mindset with regard to systems thinking, and they are successful with everyone, irrespective of age or position. These include paying attention to what people are saying without being judgmental, airing your views honestly, looking out for associations or interrelatedness in people and issues, being proactive in asking new questions and also nurturing relationships.

All in all, these strategies enhance your mental awareness and attentiveness even as they promote your creative questioning, gradually transforming your brain in a way that makes it shed rigidity so you cease to be an automated responder and become one who is thoughtful, alert and inquiring; one who is open to your own perceptions as well as those of the world.

Professor Johnson advocates mindfulness even in the workplace, because it helps to reduce, if not eliminate, stress among team members. As a result, they become more receptive to new opportunities.

You are soon going to see why mindfulness is important to everyone, and why everyone needs to examine their inner self and develop their own inner context. The inner self envisioned when considering mindfulness is one that helps you think and see in a healthy manner while making intelligent decisions. In fact, according to Professor Johnson, your survival as an individual or organization depends on how you develop your inner context.

What Mindfulness Means

Mindfulness can be summarized as having full awareness of the present moment and embracing it with openness as well as curiosity.

You can achieve mindfulness by exercising it in the form of meditation, and then you will be able to identify and acknowledge whatever challenging or painful situations you encounter, any sensations you experience, any disturbing or frightening thoughts and manage to tolerate them.

When you become good at mindfulness, you are able to control any unhealthy feelings invading your system, rendering them ineffectual as far as your perception and judgment go. Mindfulness, therefore, makes you mentally stronger and less prone to stress. This is a very welcome development considering how stressful work environments can be.

Different scholars have given varied definitions to mindfulness, but their common thread is reflected by the definition provided by Jon Kabat-Zinn, a professor of medicine, in a 2003 publication. He defines mindfulness as awareness emanating from a deliberate effort to pay attention, to be in the moment and to avoid being judgmental towards events and circumstances as they unfold.

Ruth Baer, a professor in psychology, defines mindfulness as a person's ability to consider internal, as well as external, happenings and yet not become judgmental towards them. This means you have the capacity to witness things that have the potential to trigger particular reactions from you, but you consciously make a point of remaining neutral.

Some other experts, Jean Kreistellar and Marlatt, in their 1999 writings, defined mindfulness as the ability to focus on experiences from the point of view of what is current. In short, you take things as they come, moment by moment, protecting your mind from past disappointments and pains while saving you from anxieties about the future. When you practice mindfulness, the past has little, if any, impact on you, and what the future holds does not cause you anxiety.

Mindfulness encourages a simple, practical approach to life, something that leaves your mind healthy enough to become imaginative, creative and innovative. It then becomes easy for you to develop yourself as an individual and to develop any enterprises you are involved in. However, this approach to life should not be misconstrued to mean that you should not learn from past experiences or be aware of potential risks.

It only means that when your mind is in a healthy place, you can easily set aside worries and anxieties, which then enables you to think clearly. You can then embrace critical thinking as you seek solutions to existing and potential challenges.

After practicing mindfulness and making it part of your life, you end up with elevated self-awareness and positive energy. As such, you not only become vivacious but have an extraordinary sense of peace that is usually evasive.

In this new level of self-awareness, individuals have been known to pinpoint the precise part in their system causing them ill health, which means health issues can be solved faster just by practicing mindfulness. Other people are also able to diminish pain or even clear it completely by employment of intense energy to relax and stretch out folded capillaries and nerves. This is because with mindfulness, intense and clear focus is possible.

With mindfulness, not only do you become energized but also motivated. With mindfulness, you can turn a normally boring activity into something enjoyable, and that is a good way to achieve your goals. Besides, boredom can increase incidences of stress and depression, which is not good for your personal progress or for any organization you may be part of.

Having learned the best way to attain mindfulness, it would be good to learn, too, what it is that can threaten it.

Tips to Help You Succeed in Mindfulness

- Try and maintain concentration in whatever you are doing.

- Identify and acknowledge any feelings of tension, pain or discomfort.

- Keep your memory active

- Avoid distractions when listening to someone

- Avoid dwelling too much on your future, lest you jeopardize the success of your current project or task.

- Do not allow thoughts and feelings to overwhelm you.

- Neither dwell too much on the past nor be too anxious about the future.

- Pay attention to meals as you eat and as you do other things, so that you are not always on autopilot and acting mechanically.

- When working on a task, avoid daydreaming.

- Whenever possible, address one task at a time so that you can give it full attention and concentration.

Understanding Happenings in Context

Here we shall use the word "context" to mean what is going on inside a person, the state of mind that each individual gets into, as well as the physical state where we refer to the surrounding world. Whenever people think, it is normally in the context of the physical conditions they are living in, for example, living among family members, friends and neighbors, being part of a team, an organization, a club and so on.

At the same time, it is a fact that the environment within which we live has furniture and equipment, and it also has tasks to accomplish and deadlines to beat, even as it has colleagues with whom to interact, choices to make, events to be interpreted, etc.

In these physical environments, people are interwoven with objects and events, and you cannot think of the state of one without considering the other. Therefore, even as the brain thinks of you as a person and your well-being, the other things, inevitably, feature in. That is how you get to think comprehensively about the success of a company in the context of the people in it, its physical environment, the systems it has and so on. In fact, according to Professor Johnson, daily life has the capacity to build the brain, not only on a moment-by-moment basis, but also on a continuous basis. At the end of the day, we are responsible for who we become.

For example, although each person's mind has the potential to distinguish all the sounds used in the world's languages—around 6,000 of them—it is each individual's experiences that determine what language or languages make sense to the person and which ones sound like gibberish.

It is just like the way a person's experience wires the brain for music. Early in life, the child's brain is ready to be wired for any variety or genre of music, and the kind of music the environment offers at that stage in life is what the child begins to appreciate. For those children brought up in the US, for example, their brains form circuits comprising Western musical sounds by the time they are five years old because Western music is what they are exposed to. These instances of language acquisition and music appreciation are just examples to show our capacity to sculpt the human brain.

Remember, it has been pointed out elsewhere in the book that your actions are a result of your thought process, and therefore what you make of your brain matters a lot.

According to scientists, there is a part of the brain that deals with your habits, and it comprises neuron clusters that are linked and situated close to the main part of the brain. This is known as the basal ganglia, and scientists have known for ages that it influences movement. What they did not know till later was that the basal ganglia is also responsible for the storage of habits. For example, when you do something consistently, be it at the workplace or at home, it becomes your habit, and your basal ganglia stores it as such. Lest you imagine complex habits like studying or holding regular meetings, let us cite habits such as leaving the office at a particular time, say, 4:30 p.m. every day, taking your dog for a walk every day at 6:00 a.m. and so on.

Sometimes the habits you form involve your thoughts, and so if you get used to think of depressing things, those are the ones that will keep coming back to mind, and conversely if you make a habit of thinking of constructive ideas, those are the ones that will keep resurfacing. Even in cases where you were treated with kindness as a child and were also encouraged to do acts of kindness for other people, the neurons in your habit center were wired accordingly, and the neurons formed circuits to hold that particular habit of kindness. As such, the habit of kindness remains embedded in your own brain, and it manifests itself as your personal self.

The way the brain works is, obviously, good as long as it involves good things, but unfavorable when bad habits are involved because it is very difficult to break a bad habit. Habits have their own physical space within the brain and the more you practice the habits, the more real estate they occupy.

Another reason habits are difficult to discontinue or break is that any time you attempt to eradicate them, there is a part of your brain that triggers an error message. That part is the frontal cortex; it is situated immediately above your eyes and goes all the way to the area behind the eyes. This part is essentially the brain's own error-detecting device and it keeps appraising situations on a constant basis to establish how they are faring.

In cases where expectations are not met, the error-detecting part within the frontal cortex triggers an error alert the way traffic lights flash orange to indicate pedestrians need to dash and get off the main road to the curb. In short, when time is up for pedestrians to cross the road, anyone trying to cross the road should read the orange signals to mean there is something not right at that particular time. That is the same kind of warning that goes off in the brain when it is time to perform your habitual action and you fail to do so. It is a reminder that you should be doing what you are used to doing at that particular moment as registered in your basal ganglia.

This is essentially why the way the brain works does not augur well for you or the organization you are in when you have a bad habit. When you are in your normal state and try to break a habit, your brain resists through the usual signaling, denoting the irregularity and making you feel like something is not going well. The mind, in such instances, tries to let you know that it is wrong to break the habit. Unfortunately, the error message is so strong that it can overshadow any rational thought. So even when in your rational thinking you tell yourself it is worthwhile breaking a certain habit, the brain that has been programmed to expect the actions related to the habit rebels and puts on strong opposition.

There are also emotions triggered by the brain, and they want you to pay attention to reason. Those emotions are supportive of the programmed habit and are geared towards helping the brain achieve a victory over the truth. In short, without some intervention of sorts, the brain will continue to defend its interests, irrespective of whether the habit involved has no logic or good points to support it.

Urge to Protect What is Familiar

Even after identifying systems thinking as being good for an organization, there is still the challenge of breaking old habits, including those associated with the culture of the organization. It is also unfortunate that people often act in autopilot, not thinking much about the reasons behind their actions and their implications, especially if those are things they have done over a long time. One reason is that people are simply comfortable doing what is familiar to them and often resist any attempt to change.

Research on the human brain has shown a few reasons people are so determined to preserve and protect the familiar, freeze elaborate thinking and close oneself off.

Why People Resist Change

1) The wish to belong

The brain has a longing to belong. This is the reason people are so protective of their culture and real-life experiences. They are also protective of lessons learned from personal experience, and that is because the brain happens to be a social organ and it cherishes the relationship it has with other brains. In fact, if you do not expose yourself to the culture or experiences you are used to, your brain is bound to complain.

For the brain to survive and thrive, it needs to belong because that is how it is made. That is why, even when you are resting and not engaging your brain in any particular activity, it still engages in reviewing relationships to establish how well it belongs. Thoughts of this type that cross your mind when you are not consciously directing your thinking process are of the nature of: How well did I fit? Did they find me a good fit? What did they think of me?

According to writer Michael S. Gazzaniga, who wrote on the reasons people are unique, it is the same reason people thrill their brain with gossip. It enhances their feeling of belonging.

According to Professor Johnson, the importance of belonging can be understood from the fact that people feel actual pain when they fail to belong or develop a sense of not belonging. The reason is because the feeling of not belonging makes the person feel rejected and ignored, and also mocked. Sometimes they translate that feeling as an act of being reprimanded. The pain one feels at such times is similar to the pain that is felt in instances of physical pain. So you have the different regions of the brain that respond to physical pain also responding to social pain.

The two brain regions responsible for responding to pain are the anterior cingulated cortex and the right ventral prefrontal cortex, and they respond in the same way when you are in social distress. Specifically, any time you hurt a part of your physical body, like breaking your arm, the anterior cingulated cortex instantly sends an alarm, saying pain has been caused and that is really wrong. In turn, the alarm prompts the right ventral prefrontal cortex to sooth the pain as much as it can.

The pain one feels at not belonging is so intense that individuals try their best to avoid it by conforming. One sign that someone is trying to belong is when they keep voicing other people's ideas, opinions or views earlier expressed by friends or colleagues. It is also common to go along with the cultural practices you find yourself exposed to. As such, you physically and mentally become a prisoner of context.

2) Quest for meaning

Another main reason people go out of their way to protect their convictions and their values is that the brain needs to have some meaning for it to survive. In the process of trying to find meaning, which happens perpetually, the brain seeks to identify patterns as well as order in virtually everything. Randomness, for example, bothers the mind, and it resorts to trying to seek the meaning of life. That is when certain kinds of questions fill the mind, such as: Is this job really worthwhile? Did the car crash have to happen so close to the school?

Since, in the brain's search for order, purpose and significance are encompassed in meaning, it seems to protect long-held beliefs and practices. It closes off and tries to reject any new happenings that don't fit in the patterns it is familiar with. Unfortunately, sticking with familiar patterns and protecting them is tantamount to losing out on any new meanings that a fresh context might offer.

3) Bonds of habit

Another big reason people strive to protect old thoughts, beliefs and practices is sheer habit. Expert Gipsie Ranney gives a good example of the power habits have on the brain in trying to preserve what is familiar. She points out that many CEOs hold onto the idea that performance improves when there are external incentives, yet it has been proven that such incentives actually muffle creativity and enhance conformity while discouraging risk-taking. The CEOs do not cling to the belief in external incentives because the belief matches the facts on the ground, but because the habit has been reinforced over and over again, forming a habit. Remember, the moment a thought turns into a habit due to repeated occurrence, it takes up a chunk of space in your physical brain and stays there.

How Bad Habits Can be Broken via Mirror Neurons

As has been explained, habits are hard to break because once the repeated actions have matured into habits, they occupy space in the brain and remain there, and once you fail to fulfill them as the brain expects, it raises an alarm alerting the system that something is amiss. Are people then doomed to be stuck with their bad habits for eternity? The answer is no.

When you operate in an overwhelming emotional context, with a self that is tied to habit and culture, it can feel like prison. Fortunately, there is a way out to freedom. There is a way you can open your mind, your heart and your will so that you only conform to that which is good for you and the systems to which you belong. One way out of bad habits is thinking in systems, as Peter Senge, as well as theorist Otto Scharmer, have asserted. Even if you have bad habits, there is a way you can learn good acts from other people and turn them into habits. This is helped by the fact that people are capable of paying attention, tailoring their thought processes well and looking at things with clarity.

The way through which you can learn good actions from other people uses mirror neurons. These are brain cells that enable a person to understand what it feels to be someone else.

Mirror neurons were discovered in 1996 by scientists at the University of Parma led by Giacomo Rizzolatti, as they studied the tendency of monkeys' brains to buzz with activity whenever the monkeys picked up different items. What the scientists found interesting was that the animals' brains still buzzed with activity even when they were not picking up anything themselves, but were watching their trainer pick up some nuts. In short, the animals' brains were activated as if they were doing the picking of nuts themselves as opposed to watching someone else do it.

That is the same way the human brain behaves when you watch someone else perform certain actions. The neurons in your brain fire, creating similar circuits to those that would have been created if you had been performing the actions yourself. In short, people have mirror neurons and they can help you learn and change your habits by watching other people do the things you want to learn.

One thing you may wish to note is that mirror neurons resemble all other neurons, their uniqueness being their capacity to perform two functions. When do they seem to be at work? The neurons reflect heightened activity when you perform a particular action as well as when you watch someone else perform a similar action. This is because both incidences trigger similar emotions in you, and that sends messages to the brain. It is the mirror neurons that lead you to develop the feeling you do after you have watched the action performed by someone else.

To simplify the way mirror neurons work, let us use the example of you watching a soccer match. When your team's striker kicks the ball right into the opponent's net, are you not filled with excitement as if you had personally scored that goal? How about when a penalty has been given against your team and a striker from the rival team is just about to kick the ball? Do you not hold your breath to mirror the anxiety your team players are experiencing? When you can tell the emotions someone else is experiencing, or you can literally read them on the person's face or body language, it is easy to mirror them and have the same experience the person is having.

Mirror neurons translate the correct messages to you without you having to do anything but watch the other person. Do you need to be told that someone is happy when you see them smile? In fact, not only do you mirror the person's actions through your mirror neurons, but you also understand the reason for those actions; in this case, smiling. Whether you end up with a smile on your lips or not, your brain will be smiling, and you are bound to enjoy the nice feeling of happiness the other person is enjoying.

That is why you find that one person can light up the room with their smile, while another can dampen an entire group's mood by whining about the situation, person or issue. Knowing about these mirror neurons is important when discussing systems thinking because in this approach, whatever emotions and attitudes affect one part of the system have a way of affecting other parts of the system.

Chapter 18: The Need to Consistently Think in Systems

It is important to consistently think in systems if you are to keep problems at a minimum, whether those problems are work related or just problems of a social nature. This is because, as has already been noted, systems are everywhere and everything is a system existing in an environment with other systems.

Very often, these systems have a link to one another in one way or another, although the closer the proximity of one system to another, the more their behavior is likely to affect each other. Of course, there are times when systems affect one another even when they are physically apart, like when the output of one manufacturing firm serves as the input of another.

In order to master any art perfectly, a lot of practice is required. Here we are going to consider how we can master systems thinking through both individual practice as well as collaborative learning.

How to Master Systems Thinking

1) Individual thinking

Becoming a competent and confident systems thinker begins with self-awareness and the desire to acquire the different thinking skills experienced systems thinkers usually employ. You, too, can hone your systems-thinking skills by:

- Being inquisitive

An inquisitive person will ask questions as they seek to understand the underlying problems, and then proceed to offer the best solutions. Ask questions in order to understand patterns of behavior, the feedback received, whether and why there are potential delays in feedback and also the expected consequences after the feedback. This is important in helping you understand how other people are viewing the situation.

- Learning to have a working time frame

When a problem arises, it is important to view its implications within a particular span of time and not just the present. Determine an adequate time span to study the problem at hand and then see what happens.

For example, when studying and trying to understand the rates of unemployment, you need to ask yourself what level it was at last year, how it is now and how you think it might be next year. This will help you in understanding the interconnections in the problem, which might not have been discovered or anticipated.

- Taking note of the systems around you

Note any changes around you and the impact of those changes on the system. If, for instance, a product that used to make exemplary sales in a company starts to make lower sales than usual, it is important to note what might have caused the decline in sales. Check what reinforcing processes might have been affected and how they can be improved so that sales can return to their normal level and the high sales level be sustained.

- Practicing drawing a loop

This can be done by getting stories from newspapers that can be represented in causal loop diagrams. Doing this on a weekly or even daily basis will help you master the art of causal loop diagramming.

2) Engaging in collaborative learning

Just as in other areas of learning, concepts and ideas are well understood in the company of others. In systems thinking, this can be done by:

- Finding a mentor

Finding a mentor is important because such a person guides you in understanding systems thinking better. You also can learn by becoming the person's apprentice. Mentors can also help you in the loop diagramming, as you discuss the loops with them and learn from their alternative solutions.

- Forming a learning group

The preferable group is one with people interested in learning about systems thinking, because this is where your interest lies. Engage in discussing various systems-thinking concepts. You can also decide to read a book like this one on systems thinking, and then have a lively discussion on how to apply the concepts learned. It is also important to analyze day-to-day experiences, where you can consciously seek to establish the root cause of the problem you are facing. Then you can present the scenario in a loop diagram.

How to Master the Concept of Systems Thinking

In as much as there has been tremendous development in technology and human thinking, many people still find it difficult to understand the systems-thinking concept and how it works. To fully master the concept of systems thinking, one needs to integrate several thinking ideologies.

Steps in the Systems-Thinking Process

(1) Identify the specific problem or issue at hand which you want to solve.

(2) Come up with possible hypotheses that may explain the problem.

(3) Test the hypotheses using various situations.

(4) Implement the suitable hypothesis.

If the hypothesis works efficiently in solving the problem you are facing, then go ahead and implement it. Implementation should only take place when you have fully understood the problem and you are convinced the hypothesis you are choosing is capable of working in various situations. This is important because, at times, a certain hypothesis can offer a solution to a wrong problem, in which case if the problem recurs and you try to apply that same hypothesis it will, very likely, not work as efficiently as you anticipated.

However, in case the hypothesis fails to work, then check whether the problem you have identified was the actual problem. If it was the correct problem then come up with another hypothesis and test it. Otherwise, continue your search for the real problem. To fully master the concept of systems thinking, you should practice these steps separately, one at a time, and in due course begin to combine them in different situations as they arise.

Although the systems-thinking steps are crucial if you are to succeed in optimizing the system's operations, you also need to have the appropriate systems-thinking skills. The most important of them are listed here.

Crucial Systems-Thinking Skills

- Dynamic thinking

Dynamic thinking involves identifying a problem in view of a certain pattern over a period of time. Unlike in static thinking, where people focus on a particular problem, dynamic thinking focuses on problems that have occurred time and time again, following a certain pattern. This thinking is most appropriate in systems thinking as it helps you to know after what interval a problem is likely to occur again, and what the underlying circumstances are likely to be based on the previous time the problem occurred.

- System-as-cause thinking

After understanding the interval after which a problem recurs, as well as the pattern it takes, the next step you need to take is come up with a hypothesis to try and solve the problem. The system-as-cause thinking approach helps us to come up with problems (in our hypothesis) which are within the control of people within the organization. This approach is unlike the more common approach of system-as-effect thinking, which views behavior in an organization as having been influenced by external forces and hence may lead you to come up with more hypotheses than necessary.

System-as-cause thinking is more accurate because it points out the behavior of the problem to the members of the organization, and then the members can ponder how they have contributed to those particular problems or the situations that they find themselves in. Thus, when coming up with a hypothesis, members of the organization will, most probably, come up with one that will seal the loop that ended up causing such a problem to occur.

- Forest thinking

Forest thinking is unlike the traditional approach of tree-by-tree thinking where most people assume that, to fully understand something, they must pay close attention to very specific details. The forest-thinking approach categorizes details or groups them in order to have an average picture of the organization. It dwells on similarities and not differences. For example, in as much as everybody possesses unique personal characteristics and each has varying motivating factors, each person has their own share of contribution to the success of the organization.

- Operational thinking

Operational thinking dwells on the causes of certain behavior. For instance: what is it that led to a certain behavior? This is in contrast to correlational or factor thinking which deals with predetermined factors that are presumed to have caused certain behavior. For example, in most organizations there are habits, otherwise termed "drivers," as well as factors that are critical to the success of the company. These factors are derived from correlational thinking.

In this kind of thinking (correlational), the assumption is that all factors are correlated and work together to the corresponding success of the organization. This model does not show how each factor own contributes on its to the success of the organization, but rather assumes that the factors are correlated to the success of the organization.

When factor thinking is used to analyze the learning process, it is easy to come up with a number of factors that influence the learning. These factors include motivation, class size, teacher quality, intelligence, environment, availability of teaching and learning resources and so on. On the other hand, factor thinking only shows the factors that influence the learning process, whereas in operational thinking we delve deeper into the structure of the learning process. We look at how various experiences contribute to the learning process and thus contribute to a more successful learning process.

- Closed-loop thinking

In this type of thinking we tend to view the relationship between various factors and how they affect each other in the long run. This is in contrast to straight-line thinking which views various factors independently of the effect to a company's productivity.

To master the concept of closed-loop thinking, you need to seek to understand how various factors within the system affect each other and how they bring about various effects.

- Quantitative thinking

One common assumption in the scientific world is that for one to know something, they must measure it accurately. Most business organizations are over-obsessed with measurement thinking, which emphasizes the need to get the numbers right. There are, however, variables that cannot be accurately measured such as motivation, self-esteem, commitment and devotion. Most of these variables that cannot be precisely measured are not taken into consideration when analyzing their impact on the organization's success.

However, they can be quantified in a way that helps to provide a good view of how they contributed to the overall organizational success. For example, on a scale of 0-10, where 0 represents total lack of commitment and 10 represents absolute commitment, we can quantify an individual's commitment to his or her work and see how it impacts the general performance of that individual as well as the organization at large.

- Scientific thinking

In this type of thinking, we acknowledge that for there to be progress in science, the existing falsehoods within various hypotheses must be discarded, and that can be done by adopting more acceptable hypotheses. Most business leaders make unrelenting efforts to defend their business models by touting their previous track records. In systems thinking, we try not to fully validate our models but instead check what falsehoods might be buried within those models.

The assumption is that every model is wrong, and we therefore seek to establish what is wrong in them. Numerous tests are done on the models until we identify where there is an error within the system. We also seek to establish under what circumstances the error arose.

To master the concept of systems thinking, it is imperative that you comprehend each of the above thinking skills and practice them each, one at a time. When you have adequately mastered the seven thinking skills, you can engage all of them as a systems thinker and have your system operating optimally.

Traditional Skills	Skills of Systems Thinking	Stage
Static Thinking People tend to focus on specific events.	Dynamic Thinking People tend to articulate the problem in terms of behavior pattern over a reasonable period of time.	Identifying the problem or issue of concern
Seeing the system as reflecting effect People look at something happening to the system and begin to think of external factors that apparently caused it.	Seeing the system as being the cause People look at something happening to the system and begin to think of possible causes within the system itself. Responsibility is placed squarely on managers of system policies as well as those charged with monitoring the system on a daily basis.	Identifying the problem or issue of concern
Employing tree-by-tree thinking Trying to delve into details in the quest to know what is happening within	Employing forest thinking Trying to understand everything that is happening in the system, because even	Constructing relevant hypotheses

the system.	without the details you will understand the relationships that exist within the system.	
Thinking in terms of factors

Noting factors that contribute to the result or those that have an impact on it in any way. | Thinking in terms of operations

People concentrate on establishing causality and trying to learn how a particular behavior comes into being. | Constructing relevant hypotheses |
| Thinking the straight-line way

People consider causality as flowing only in one direction, and one cause being independent of any other cause. | Thinking the closed-loop way

Taking causality to be a continuing process, as opposed to being a one-off event. This process is seen as having the effect of feeding back to the system in a manner that influences the causes as the causes themselves continue to affect one another. | Constructing relevant hypotheses |
| Adopting measurement thinking

People search for data that has been | Adopting quantitative thinking

Appreciating the possibility of being able to quantify | Constructing relevant hypotheses |

measured with precision.	without necessarily having precise measurements.	
Thinking in terms of proving-truth People try to prove that the models used are correct by validating them using historical data.	Thinking scientifically People appreciate that every model is a working hypothesis and its applicability is limited.	Testing hypotheses

The Importance of a Management Theory

Due to increased competition between various organizations and the high expectations by customers to get the best quality of goods and services, managers of various institutions are faced with overwhelmingly high expectations. Faced with these challenges, they turn to new methods and approaches of solving problems which offer quick solutions to their challenges. However, the problems reemerge because the solutions were only temporary. The organization is then engaged in a continuous cycle of solving problems which later resurface.

In cases where the solution may seem permanent, sometimes the managers have no idea how the solution came about. In short, they cannot explain the procedure or the process to anyone with a similar problem in a way that will help solve the problem. By the same token, the organization cannot also benefit from that successful experience any time in the future.

The nature of relationships in an organization influences the thinking and actions of people in the organization, and this also determines the quality of results to be expected.

Why Most Approaches Fail

Most theories and approaches fail because organizations tend to copy other theories that have been found to have been successful in other institutions. Managers forget that the ethos and situations in organization A are different from those of organization B. Thus, the theories or approaches employed by organization A must not necessarily offer the same results in organization B.

Most managers also tend to look at individual factors and how they may have contributed to the overall success or lack of it. What they forget is that the success of the organization is as a result of the success of individual interrelated departments in the organization. Companies often tend to focus their energies on departments that they think are key to their success.

Take, for instance, company X which deals with cookies. After realizing that their sales have slowed, they set up a special team to market their products. The managers are also instructed to focus their energy and resources on marketing. The approach works and the company increases its sales volumes after a short period. Nevertheless, the success is short-lived. This is because other departments, like manufacturing, were not prepared to cope with the high demand of cookies. Had the management sought to understand how the various departments are interrelated and offered to support all of them accordingly, the manufacturing department would not have been overwhelmed and the success would have been long-term.

Importance of a Customized Approach

Most companies tend to associate theories with academic circles. It is, however, important for organizations to have customized theories, approaches and philosophies in running their businesses. Theories offer knowledge that is applicable in various circumstances including in the corporate world. They help us in predicting behavior or even explaining certain circumstances.

A customized theory or approach in an organization helps to predict the outcome in a particular situation. It also helps members understand the organization better and hence improves its performance capacity. Managers who clearly understand the theories of success in their organizations will, most likely, tend to invest more in areas that are more critical to that success.

In systems thinking, we not only identify individual factors that lead to success but how the individual factors are linked to the overall success of the institution. This helps in creating a chain of factors that leads to the success of the organization. It also shifts emphasis from individual factors to the interrelationships between these factors.

Theory as an Assessment Guide

It is important to continually assess a company's theory in order to know whether it is contributing positively or negatively to the institution's success. Let us have a sample of a successful theory in Company X.

We can base our theory on the idea that workmates who relate well are most likely to show respect, trust each other and work as a team. This results in quality thinking and appropriate actions taken in favor of the organization. This, in turn, gives good results to the organization. When people achieve more as a group, they are motivated to continue achieving even better and better results.

In such an approach, all factors are interrelated and, where one is not functioning properly then it should be reviewed to ensure smooth running.

In most organizations we have a top down method of achieving results. If the results are not as per the expectations, the top leadership will offer "help" to the bottom level. This may offer quick and short-term results but the "help" may be an unintended hindrance which may obstruct beneficial positive actions in future. The so-called help may create mistrust and low morale among the lower-level employees, which will then lead to poor relationships and, ultimately, poor results for the organization.

Managers who do not understand their success theories may start celebrating and indicating that the "help" offered produced results. What they will never know is the long-term impact of that help. If a similar situation arises in future, the leadership may offer the same help again but the earlier results might not be replicated. By having a theory, we can clearly see the negative impact of the help, and we can come up with ways to counter the negative effect it had on the organization.

Managers' Roles in Success Theories

Managers should come up with theories that develop people's full potential, as such theories lead to job satisfaction and more commitment. This will, at the end of the day, bring better results to the organization. Managers should also ensure that their efforts to achieve better results do not end up becoming a hindrance to the quality of relationships, but are accelerators to even better relationships.

Quality relationships lead to quality thinking, and, of course, quality actions follow, which then bring about the desired results. If good results are achieved, the managers are likely to be under pressure to improve on them going forward, and hence they will end up improving the quality of relationships in the organization. This will then lead to a continuing cycle of success for an indefinite period.

Theory building is a fundamental role played by managers in ensuring the smooth running of the organization and the achievement of sustainable long-term results.

There are some attributes that are crucial for a manager to be able to successfully implement a theory, and they are listed here.

<u>Qualities Required of a Systems-Thinking Manager</u>
- Improved communication skills

This involves how the manager communicates requests, orders, commands, suggestions, compliments, complaints, instructions and inquiries. The manager's communication needs to be as clear as possible. He should also create an environment where members can dialogue freely and give their perspectives without fear of intimidation. This creates respect and trust between the manager and the rest of the staff, which is then reciprocated in the form of quality relationships.

- Self-reflection

Managers need to self-reflect so they can acknowledge to themselves the attributes that may hinder their attainment of desired results. They can then make a conscious decision to change those negative attributes, and this will lead to quality thinking.

- Systems thinking

Managers need to have a clear understanding of the different departments in the organization and how they are interrelated. This will help them understand the effect of an action taken within a particular department on another department.

- Commitment

Managers need to have a clear picture of their own personal mission and the organizations' vision, both of which should be clearly spelled out to the other members of the organization. This will be helpful in that people will be able to identify with the vision, and that is likely to make them more committed to working towards attainment of that vision.

Chapter 19: Manufacturing Company Leads via Systems Thinking

In this chapter we are going to review a modern case where a manufacturing and trading company became a global leader by following the systems-thinking approach and sustained its enviable growth for around five decades. However, the global financial crisis came, the company management succumbed to pressure and the company fell into a crisis. The problems escalated simply because the company management switched to a management approach that ignored systems thinking. This is the case of Toyota, which has been explained very well by Professor H. Thomas Johnson, a renowned management thinker whose specialty is sustainable business practices.

Whenever he gives talks to eager audiences, he likes to explain the mistakes people make in the way they view business, and then explain the correct way to view the same. The listeners can generally relate to his earlier mistakes, as his erroneous perceptions happened when he was a young professional accountant and continued to the time he served in the academic world at the university where he taught economics and management accounting. So what was Professor Johnson's mistake?

He looked at businesses from a standpoint of financial information, which mainly comprises market prices, cost reports and accounting statements. As you can see, there were no relationships in his thinking at that time. Then it happened that, many years later, he was introduced to Toyota, and a look at how the company operated helped him to learn that enhanced financial information, even when it included valued activities such as budgeting, is not sufficient to help an organization's performance to soar. He learned that Toyota's exemplary performance was a result of the way the company uniquely organized relationships among its employees as opposed to pushing them to perform by giving them financial targets.

Incidentally, as he was learning how Toyota worked, which happened by chance, he was also doing research on modern science with a special emphasis on life sciences, astrophysical cosmology and some other areas to do with people. Soon he found a close relationship between the company's style of operating and the way different earth systems are particularly organized by nature to be interdependent and supportive of one another's survival.

That became Professor Johnson's epiphany—being able to find similarities between nature's processes that ensure sustainable growth of natural systems and Toyota's management and operational practices that supported robust sustainability.

Financial Crunch Confuses Analysts

However, when the financial global economic crisis of the 1990s caught up with Toyota, after the company's way of working had kept it in the lead for 50 years, it was now facing product recalls and other financial problems. Some analysts began to question the support Professor Johnson had expressed of Toyota's way of management and the way they ran their operations. They thought the company's management model was not good after all. However, the professor stood his ground and explained that it was not systems thinking that was in question in this case, but the move to deviate from it in an apparent panic.

Before 2000, Toyota operated based on a systems-thinking approach and became a world leader in car manufacturing, giving established companies like Ford a run for their money. However, when the company began to feel the effects of the global financial crisis that emanated from the fall of the Lehmann Brothers, Toyota's top management panicked. They began to seek quick fixes to problems that were complex; problems that went beyond Toyota.

All those years before 2000, Toyota had succeeded because the company's unique thinking defied the traditional approach of cause-and-effect and instead sought to strengthen company systems. Those stunning practices are the ones that produced the company's enviable results and drove its sustained growth for several decades. What the top management ceased to appreciate when faced with the sudden financial problem was that the thinking that had seen the company succeed over the years would still help the company pick up progress once the overwhelming global crisis was contained. As Professor Johnson has noted, the management's thinking had, all along, implicitly anchored the company's systems, and hence its operations, to the real world's natural systems.

Failing to realize the potential danger of ignoring that reality, the company's management began to address its problems with a traditional approach. Toyota began to rely on financial abstractions, which only provided virtual reality. This was something the professor understood very well, having begun his career as an accountant and worked at the university in a business-related field. He could see how limited the thinking later adopted by Toyota was, whereas Toyota's management probably thought it was good just because other companies were using it. In the worry of the moment, they ignored the sustainability of growth their earlier thinking had provided.

Why Big Corporations Fail to Work Optimally

It is typical of big corporations to put a lot of emphasis on the virtual reality that is based on financial abstractions, and this is the same thing financial markets do. In the meantime, they become indifferent to the solid reality of other systems, both human and nonhuman. For that reason, it has generally been assumed that fluctuations in financial fortunes year after year, or after every couple of years, is normal, which should not necessarily be the case. This outlook contributed to the global financial crisis of the 1990s that heavily hit Toyota, destabilizing its financial performance.

Toyota's Founder Acknowledges Deviation from Fundamentals

Professor Johnson was vindicated when, in 2009, Toyota's founder, Shoichiro Toyoda, who is also Toyota Motors' honorary chairman, announced the company was ready to go back to its age-old management approach—the one similar to the universe's natural systems. He noted that the global recession had not been the main culprit for Toyota's failing fortunes. That had been the way the top management had handled operations in the years around 2000. They had tried to accelerate company growth by pursuing finance-based growth, while ignoring the fundamentals on which the company had developed over the years. The octogenarian criticized Toyota's top management and instituted major changes in the make-up of the top management team.

Toyota Publication Explains Company's "True North"

Toyota's success has been based on the principles of systems thinking, although within the company they see it as focusing on what they term "true north." In short, they do not just look at the number of cars being churned out to the market, but at the working of their internal systems that include workers and their attitudes.

However, it is apparent that Western analysts failed to understand Toyota's management approach when, in the 1980s, they termed it "lean." It is true one could see characteristics of the lean approach in some branches of Toyota at the time, but those that Westerners observed in the company's plants like Kanban and Jidoka were only temporary fixes and countermeasures that the company appreciated as such. In fact, the company has never purported to use the lean management approach.

They are always conscious that the company's fundamentals lie elsewhere—strengthening all systems, no matter how small, and seeing them work in tandem with one another. In fact, there is a 2010 publication detailing Toyota's problem-solving approach which is unique, and which shows the company management's fundamental thinking in systems.

How Systems Thinking Distinguishes a Company

In the case of Toyota, with its systems-thinking approach encompassed in the company's "true north" pursuit, managers and workers pay attention to generating well thought-out processes and sustaining them on a continual basis. The result is a process with the capacity to produce results that are enough to sustain Toyota's daily activities indefinitely. It is worthwhile noting that the company focus is on well-sustained processes as opposed to set revenue targets.

If revenues drop while the company's processes are working optimally, the company will need to check whether there are external systems whose relationships have changed, since the company does not exist in a mutually exclusive environment with other systems. Such a company distinguishes itself from others in its focus, which is what guides its manner of management. When its focus is on enhancing processes for sustainable efficiency, other companies focus on the bottom-line in the immediate future. They push everyone to focus on producing the maximum possible yields within a given short period. Moreover, they do not set those targets from internal evaluations of their capacity, but are rather influenced by global financial trends for the most part.

Companies that do not employ systems thinking, either consciously or unconsciously, handle their departments, plants and systems as independent parts of the company, whose output or contribution is mechanically assembled to produce results for the company. In this scenario, every part of the company is taken to be an isolated entity, which can be manipulated in order to see predictable results or consequences. This is not the way a company like Toyota views its plants and other systems within it.

Toyota views its workers and systems as comprising a web of relationships which are nurtured in patience to enable them to produce results within a process that is both complex and nonlinear. In short, nothing within Toyota is an independent component. Everything is an important part of the process-based complex system. That is why

Professor Johnson equates Toyota to natural living systems with their inherent interrelationships.

In short, during Toyota's peak performance years, the company was process-driven as opposed to being results-oriented. When other companies were being influenced by the behavior of financial markets in their operations, Toyota stuck to its management approach and maintained its operations as it normally had. As a result, the company continued to report impressive performance every quarter of the year, regardless of what the financial markets looked like or predicted. This was because process-based systems have progressive growth that is sustainable. Toyota paid attention to systems processes, meaning its management essentially employed the systems-thinking approach, and that led to several decades of exemplary financial success. As another expert, W. Edwards Deming, advised years ago, if you want your processes to produce great results, avoid demanding that people meet specific targets by hook or crook.

One additional takeaway from the case of Toyota, besides that of underlining the importance of thinking in systems, is the fact that a company uses the systems-thinking approach does not make it immune to external pressures, especially when they are of the magnitude of the global financial crunch of the 1990s and early 2000s. However, if a company is confident about its processes, management need not succumb to the pressure of financial markets, which, in any case, are always fluctuating. Even if the company finds it necessary to take some short-term remedial action, the focus should not deviate from enhancement of company processes, because that is the approach whose results are sustainable.

Chapter 20: Familiar Business Scenarios Requiring Systems Thinking

Sometimes when a concept is reviewed and given publicity by renowned academicians and high-level researchers, people tend to think the idea is mainly academic. Nevertheless, as the case of Toyota shows, where the company excelled in the principles of systems thinking without necessarily defining its approach as such, there are many business scenarios where the enterprise could alleviate many problems just by adopting the principles of systems thinking. Of course, for any system to succeed it is important to first analyze the prevailing trends and then to plan accordingly.

Systems Thinking in Strategic Planning

Systems thinking is by far the most useful tool that a business can use for proper strategy analysis. The good news is that you do not have to run your business like a computer program. In fact, it is not a prerequisite to be good at computers for you to succeed in strategy analysis, although it is an added advantage if you are conversant with technology. Systems thinking can be applied by anyone since it is a way of reviewing and analyzing a system by thinking through the various connections that affect the business and demonstrating how those connections all work together within the system.

The principle idea behind systems thinking is that we can understand what would otherwise be a complicated system by studying how the different pieces work interconnectedly to change the behavior of the system. To understand this better, here is an example of a restaurant setting.

If you go to a restaurant, you will, very likely, encounter one of these two scenarios: you will either get a seat right away or you may have to wait until a table is vacant. This is attributed to a fairly simple system. There is a limited pool of resources, in this case, tables. The availability of this resource has the capacity to turn customers from one state, one of waiting for a table, to another state—one of being seated at a table. It is an easy process that takes little time if the tables happen to be vacant, because all you have to do is speak to the host and you are soon seated at an empty table.

The process, however, takes more time if the seats are all occupied, because you have to wait for the seated occupants to leave and for the attendants to set the tables for the next clients.

Several variables and processes determine whether or not you will wait for a table to be available. The first variable is the number of tables at the restaurant. It is easier to fill fewer tables than it is to fill many tables. If you have 500 tables in your restaurant, you will find that it takes a longer time for people to fill them than if you had five tables. In the latter case, the restaurant will be full to capacity as soon as the fifth set of guests arrives. Another determinant is the rate at which new guests arrive in contrast to the rate at which the served guests leave the tables. When guests arrive faster than the current guests can leave, then there is an imbalance and consequently, the restaurant will have an overflow. The waiting customers will form a waiting line that gradually grows as more customers flock in.

Yet another major determinant is how fast guests finish their meals and leave the tables for other occupants to use. It is in the interest of the restaurant owner to have the guests leave as soon as they are done with their meals, especially during busy hours. Sadly, most customers hang around after their meals, causing the restaurant to fill up. This is a challenge shared by most restaurant owners across the globe. A business can turn its fortunes around by addressing one process within the system, as is the case with the following restaurant whereby addressing the seat-filling issue led to efficiency and profitability.

Benihana, an American restaurant that owns many franchises all over the world, came up with an ingenious way to solve this problem. The guests are treated to an entertaining show from start to end as a way of preparing and serving meals. The guests then leave shortly after the chef is done. This system allows the restaurants to turn their tables more quickly, and the hastened process has generated good returns for the restaurant and its franchise.

Another way of curtailing long lines is through raising prices. Busy hours with a higher price on meals guarantee good profits while still ensuring there are no long lines from waiting customers. The downside of this approach is that raising the prices too high can lead to empty tables. It is therefore important that the manager strike a balance between too affordable and too high.

Oftentimes, customers ask for expensive and difficult services from companies as part of customer support. These demands can take a toll on a company if it offers all after-delivery services for free. The company bearing the cost of service finds that, in due course, it begins to make less and less profits. In addition to that, there is an increase in demand for more of those services just because they are free. To discourage excessive demand for after-delivery support services, you need to charge a fee for those services. Putting a price on the services not only creates a balance between supply and demand but also ensures that only the customers with genuine serious needs are served.

There are situations, however, where putting a price on those services in a bid to control demand may not solve the problem. A proper scenario is where fellow competitors offer their services at a lower price or for free. In this situation, you must find a different set of strategic responses if you want to retain your share of the market.

Chapter 21: Systems-Thinking Principles Fit for The Health-Care System

It would be helpful to adopt the systems-thinking approach in a quickly changing health-care system. This would help manufacturers in the medical field to develop new policies and products that match the changing clientele. Often, failure is a result of delays in responding to the changes that occur in customer demands. Systems thinking-approaches can alleviate such challenges by ensuring systems are well streamlined and enhanced to work with other systems run by different stakeholders.

Why the Systems-Thinking Approach is the Way to Go

It is important to use systems thinking because:

• In this approach, rather than taking a problem and breaking it down into smaller constituent parts, we seek to understand the relationships between problems so as to come up with a comprehensive solution for them.

• Systems thinking helps us to understand how a solution to one particular problem impacts the whole system.

• It builds a bigger picture of the problem as well as the anticipated impact of the solution.

• It addresses recurring problems that might not have been solved using other approaches.

• It creates a common approach to the problems identified and the solutions to be implemented.

Major Factors Addressed by Systems Thinking

The following are basic factors that systems thinking addresses:

- Purposefulness. This involves not only understanding why customers choose your product but also why they use it in solving various problems.

- Stability. This involves the ability to strike a balance between various needs and variables. It is also important to anticipate the change in relationship between your product and other products in the market.

- Relationships. It is important to understand how your product is related to other products in the market, and the systems-thinking approach helps accomplish this. Understanding relationships encompasses understanding how the external environment affects your products, an aspect that is also covered in systems thinking.

- Consequences. Sometimes solutions may not always offer the intended results and may have the opposite results. For example, to reduce the cost on support staff, a health information company included the ability for users to customize and self-configure their personal devices. Unfortunately, the operation of the device became so complicated to the customers that the company had to employ more staff to train each customer on how to operate the devices.

Luckily, when using the systems-thinking approach, you have room to analyze many of these possibilities in advance, as you are continually in communication with the different stakeholders.

- Emerging issues. Sometimes a solution can be found and implemented, and it ends up offering a solution, not only to the initial problem but to another problem that was not part of it.

Whereas other approaches will see that as an added advantage and nothing more, in systems thinking, the people involved in the affected processes will do more analysis to try and establish whether the apparent solution is real or just a temporary incidental fix to the second problem. That is a great way to preempt future problems, because if the second problem was not solved from the core, it would, very likely, take the organization by surprise when it cropped up later, and possibly in a more serious way.

Practical Challenges of Systems Thinking and Modeling in Public Health

Information from the US National Library of Medicine shows the government considers systems thinking as a viable approach. There is information that is more than ten years old indicating that relevant health partners in the country have been conducting research to see how best to implement systems thinking in the health sector. For a system which, for much of its life, has used traditional approaches to management, it is not surprising that challenges have been identified in the various analyses and testing that has taken place. Fortunately, the researchers did not stop at identifying challenges, but sought to seek ways of handling those challenges so that the government and its citizens can enjoy the benefits that come with thinking in systems.

First of all, the parties involved could not ignore the fact that systems thinking was gaining currency among the knowledgeable proportion of the populace, as well as among various leaders in the health sector. They could imagine a health sector whose enhanced processes could render high quality services in a sustainable way. However, they also acknowledged that introduction and implementation of systems thinking was bound to encounter challenges and that the biggest priority would be to identify what those challenges were, particularly from the perspective of the health professionals on the ground; those practicing within the public sector.

The next matter of importance was to lay out a method of study, where concept mapping would be done and more than 100 participants were involved. At the end of the day, it was established that there were as many as 100 challenges facing implementation of systems thinking within the complex public health sector. However, the researchers found a way of grouping those challenges into eight categories, following their respective dynamic interactions.

Ultimately, guidelines involving a mere eight rules were drawn up, and those rules were crafted with the eight grouping challenges in mind. Thus, the researchers laid groundwork for improvement of public health systems for the welfare of the public.

One undeniable challenge is the complexity of players and systems within the modern health sector. They include government entities that also happen to have variations because some are local, others regional, others national and even others international. There is also a conglomeration of nongovernmental organizations involved, which include foundations, coalitions, partnerships, special interest groups, advocacy groups, profit-making medical systems, nonprofit medical systems, businesses and then the public as a whole.

The complexities not only arise because of the wide range of players involved in the health sector, but also because of the wide range of medical challenges involved. These range from simple and easy-to-treat infections to serious health conditions requiring long-term management. Some of these medical conditions not only pose a challenge to the health sector but also other sectors that involve social interactions and workforce management. Cases of obesity, tobacco use and serious infectious diseases are just some of those that pose a great challenge in the health sector, and they affect other sectors as well. This complexity is one more reason why the health sector should insist not only on adopting systems thinking, but also why all the players involved should do the best they can to enhance the systems involved even as they embrace the process-driven approach.

The lead players within the health sector acknowledged that systems thinking was already being embraced in other sectors, and although it was seen as an intellectual endeavor, they were ready to embrace it because it promised not only better service delivery but also high-quality service in a sustainable manner.

First of all, in order to understand the context within which systems thinking is being discussed here, it is important to understand what comprises a health-care system. This system can be aptly understood as a set of interdependent agents, which includes patients and their caregivers, and which are linked by a common purpose as well as their practical knowledge. It is also important to appreciate the complexity of the health-care system, which basically emanates from that dense interconnectedness of smaller health- care systems.

One cannot consider systems thinking within the health sector without factoring in models such as those involved with ecology, public health practitioners, population health, public health and others, and that wide involvement is what makes implementation of systems thinking in the health sector a significant but worthwhile challenge.

Take the example of the issue of tobacco control. Whereas it is primarily seen as a public health issue, one can only appreciate its implications and its challenges by examining the other systems affected and understanding how dynamic and complex the interrelations of those systems is. Reports from medical personnel on the effect of smoking, such as the 1964 US Surgeon General's Report, had, and continue to have, an influence on policy. There is also the marketing angle, with various firms seeing the negative impact curtailing tobacco advertisements will have on their bottom-line. Tobacco farmers have their lobbyists trying to influence policy, and there are also private medical insurance firms putting in their word.

This complexity that is seen in times of policy making and implementation is similar to what the health sector faces when trying to change health-related systems from traditional management approaches in favor of systems thinking.

It may be, for instance, that the surgeon general's report played a crucial role in instigating the public policy environment, which, in turn, catalyzed litigation that then led to the Tobacco Settlement Agreement many decades down the line. This is likely the series of actions begun by the report that led to many states raising taxes on cigarettes and to stringent restrictions on smoking within public areas. The same report may have led to restrictions on tobacco advertising.

Although the report is credited with having influenced many positive actions, it is also deemed to have been responsible for many negative consequences that were not anticipated. Examples of such negative responses include the formation of strong lobby groups working for the tobacco industry, covert attempts to undermine any efforts being made to conduct research on tobacco and its effects and others.

Clearly, a single action can have a wide range of effects, some of them positive and others negative. Some of the effects, while seemingly helpful, may actually be as bad as the initial problem, if not worse. In the case of tobacco, where advertisement of cigarettes was widely banned, shrewd entrepreneurs came up with variations of cigarettes purporting to offer smokers a less harmful product. However, what cigarettes termed as light cigarettes were not much better in terms of people's health. Although those cigarettes termed light are said to have lower tar and nicotine levels, they have also been found to be responsible for a type of cancer that is even more lethal than the one mostly associated with normal cigarettes.

Another consequence that ended up watering down the gains made by the report was the course cigarette advertising took. Since restrictions were put on advertising so that cigarette companies could not advertise their products on billboards, the companies sought to beat the legal restrictions by advertising their products at retail stores. Billboard advertising was banned through the Master Settlement Agreement. With cigarette images flooding the scene at almost every retail store, the mode of advertising, though not sophisticated, reached a wider consumer base, including those most vulnerable—youth and children.

Everything that anyone does—you, the company you work for, your neighbors—is interrelated with other systems, and unless you have done a reasonable analysis to anticipate how the other players are going to respond to your actions, you may have a rude shock. That is why, although systems thinking is not a panacea to all the problems that ail the world, at least it is close to being the best approach towards management that is geared towards sustainable growth and development.

Chapter 22: Systems Orientation – The 5 Cs of Systems Thinking

One of the most important reasons systems thinking differs from many other problem-solving approaches is that it factors in the fact that whoever is trying to solve the problem is part of that system. Even when you are analyzing faulty processes at your workplace, it does not matter whether the faulty point falls under the maintenance of another department. You are still part of that faulty system. After all, as has been explained before, the world is one big system with many other systems within it, some visible and others not.

For that reason, it is imperative that you learn a unique way of viewing the world objectively if you want to become an accomplished systems thinker. This is because ordinarily it is very difficult to objectively influence a system you are a part of. In fact, it is one of the reasons attempts to problem-solve often fail to bring about long-term solutions. Even when the solutions seem tangible or solid, they still recur after a while and, when they do not, other incidental problems arise, whose emergence is pretty difficult to comprehend. This endeavor to understand systems in a unique way is referred to as systems orientation, and it not only enhances our capacity to use the appropriate problem-solving tools, but it also becomes stronger as we regularly engage the right tools to solve problems. Which, then, are these systems orientation tools?

The Makeup of Systems Orientation

Systems orientation can aptly be summarized as follows:

(1) Curiosity

(2) Clarity

(3) Compassion

(4) Choice

(5) Courage

The five attributes are popularly termed "The 5 Cs Of Systems Thinking." What do they really mean in the context of systems thinking?

How the Five Cs Work in Practice

(1) Curiosity

This is not the curiosity where you try to peep through the fence to see what your neighbor is up to. This is different. It refers to what you do in a situation where you are trying to accomplish something and things do not seem to be working as you anticipated. You may have been working on a project, and as far as you were concerned, everything was going fine. Then, as you were nearing the end and trying to wind up the project, you realized the result it was pointing towards was nothing like what you expected. What did you do? That is what curiosity is in systems thinking, and it means take a step back.

Such advice might be counterintuitive, considering many observers would urge you to try harder, but it makes a lot of sense if you think about it critically. Why would you want to try harder if you have no new strategy or new information to show you a different angle to the problem? So, in systems thinking you halt and generally take stock of where you began with the problem.

What did you see in order to label the problem you are tackling? What other system is the so-called problem associated with? Might the real problem be emanating from there, so that this is only its symptom? Was there another system problem solved recently, because this is probably an incidental problem created by solving that other issue? In short, curiosity is appreciating that you may not have understood the problem properly and stepping back from trying to solve it with a view toward inquiring about it in more depth. When you are curious, it means you are also willing to acknowledge that the suggested solution may not be the right one.

Often when people do not step back either to get a better perspective to the problem or to learn more about it, they will do things in futility, like:

- trying to tackle problem symptoms head on

Some people think when a problem has been identified, such as declining sales, the best solution is trying to generate more sales. That is when you see a company reducing prices, putting products on discount for no other reason but to increase sales.

- framing the so-called problem in terms of the presumed solutions

You might hear a manager talk of the problem being to determine how to bring down prices. Can that really be the core problem?

- generating more policies

Is generating more policies more of a distraction than a solution? In fact, in situations where what was presumed to be a fitting solution is not working, formulating additional policies does not help the situation. Often it exacerbates the problem.

(2) Clarity

Clarity in systems thinking is what is born out of curiosity. When you step back and try to look at the system with fresh eyes, as in the state of curiosity, this is when you are able to dig deeper into the system by making inquiries. Hence you are able to see the problematic situation with more clarity. This time around you are bound to be more accurate in your observations, and you are likely to be more comprehensive. After all, you will have discovered that too much haste can easily lead you to a dead end. So, at this juncture you are likely to look at the problem more exhaustively.

Clarity entails embracing other mental models and being prepared to use them to solve the problem at hand. In addition, it includes appreciating the possibility that you could be part of the problem, and therefore must look at ways you might have inadvertently contributed to it or even fueled it.

When you devote sufficient time to establishing what the real problem is, you are likely to come up with lasting solutions. When you do not spare sufficient time and attention to trying to find out what the real problem is, what you term solutions are normally kneejerk solutions, the cause-and-effect kind. At the same time, hastening to solve a problem before making certain you have identified the real issue can be extremely costly in terms of time and money.

(3) Compassion

Compassion is basically born out of enhanced clarity. It is the capacity to appreciate that everyone is part of the one common system, and that being the case, no single person is solely responsible for blame. If blame has to be invoked then it is everyone in the system who needs to take the blame in a collective sense. However, since blame never helps to solve a problem, everyone is responsible for ensuring insights into the problematic situation are shared, as well as viable alternatives to solving the problem.

In essence, compassion is what drives people away from thinking of apportioning blame, and instead pulls people towards taking responsibility. That way, it is easy for every individual in the picture to see how they may have contributed to the problem, either by action or omission. Sometimes it may dawn on you that you are responsible by way of the thought processes you adopted in the situation, group policies or real actions.

One of the greatest benefits of compassion is that it gives you and the team power. You have the power to acknowledge what happened and you have power to influence the next course of action by choosing your thought process and actions going forward. In short, when you have compassion toward someone or something you are bound to take responsibility, so you have power to influence what happens to that person or thing because of that direct responsibility.

Take an instance where you acknowledge that a good degree of the pressure the group is receiving from the company's top management is a result of your failure to have regular communication with the senior managers. Is it not in your power to take action towards remedying that? In this case you would henceforth make a point of keeping the senior managers up to date with whatever you are doing, with a view to relieving the pressure coming from the top.

(4) Choice

Choice in systems thinking refers to the capacity to appreciate the need to address a problem from different points or angles, perhaps finding a multi-pronged solution. Once you do that you are unlikely to chase symptoms as the real problem escalates, and you are also unlikely to have other areas of the system suffering side effects from your chosen solution.

When people fail to think in systems, they often come up with a single solution just because they identify the problem as one. What they forget is that the single problem has a negative effect on more than one system or parts, whether that is visible or not at that point in time. In such cases, if you are part of a company, you are likely to hear suggestions indicating the need to go for a market-driven approach or a more technology-driven approach. Who said solutions to a problem need to be mutually exclusive?

Luckily with choice, you have the leeway to implement more than one solution to work concurrently. You could, for instance, make use of the market-driven approach while you consciously implement a technology-driven approach over a period of time. That way you will still increase your market share as you satisfy your technology-savvy customers. On the other hand, if your two choices are either to increase product prices or to reduce production, you might opt to use both approaches to solving the problem you have identified. You can raise the product price a little even as you reduce the availability.

As you make these choices, it is bound to dawn on you that every move you make is going to have an impact, whether implicit or explicit. This is one fact that makes critical thinking inevitable in systems thinking. You must be aware of the potential consequences of your actions. Nevertheless, knowing the possible consequences does not necessarily mean you are going to back off implementing what you deem to be the best multi-pronged solution, rather you are prepared to handle the consequences in a manner that ensures they do not destabilize any system processes.

(5) Courage

Courage in systems thinking refers to the capacity to appreciate that, even with the range of alternatives you have to help you solve the problem at hand, including some that are entirely fresh to the organization or whatever other system you are dealing with, it is not obvious that the solutions will be well received by the rest of the team.

The reason is that people are used to seeing old solutions modified or repeated. Other times, people prefer quick fixes because they seem to produce instant tangible results. What such people fail to appreciate is that the tendency to employ the same solution over and over again, or modified versions of the old solution, is responsible for the recurrence of the problem, system behavior that ends up costing the organization more than is necessary in terms of resources and downtime. Quick fixes are like bandages, which, unless immediately followed up with a solid solution, do not provide a long-term solution to a problem.

Through courage, systems thinking enables you to take a strong stand whenever there is a problem to be solved, and you have confidence in the position you take because it is backed by comprehensive and systemic analysis of the situation. Hence, your courage is geared towards supporting solutions that are sustainable, freeing the organization of the problem and its possible offshoots once and for all. You can also aptly credit courage with enhancing your curiosity as well as clarity, because it is courage that gets you to acknowledge what you do not know, along with the accepting the probability that you are somehow responsible for the problematic situation that has developed.

All in all, you can sum up systems orientation as a predilection that causes you to identify alternative manners of thinking, behaving and acting. Once you have become an accomplished systems thinker through regular practice, you will find yourself making an impact in the world around you with greater effectiveness, and your personal life will also be positively impacted as you view the world by way of a systemic lens.

The Systems Thinker in a Nutshell

As an accomplished systems thinker, you will find yourself inquiring about problems with more depth and appreciating your own responsibility with more clarity. At the same time, you will be able to face resistance posed by others with more compassion than you would otherwise have done—resistance that, as a systems thinker, you will have anticipated, anyway.

You will also be in a position to handle the entire problem-solving process with confidence and all the while be creative in the search and implementation of a viable solution. You will become a person who has the capacity to take a firm stance in favor of what you believe to be right, viable and sustainable.

Chapter 23: The Role of Systems Thinking in Education

Have you been looking at systems thinking as a problem-solving approach for business enterprises? The truth is the approach does help to solve many problems in the business world, but it is by no means only applicable to that area. You only need to remember that systems include everything that exists, both live and inanimate, and so the approach will be just as helpful in other areas as it is in the business world.

In this chapter, you are going to learn how systems thinking can help alleviate all manner of problems, from administrative matters to student performance and a lot more. It is important to remember that people who run businesses and influence politics in Congress have all passed through the education system, and once their tenure is done another generation will take over that is currently within the education system or which will have just left it. It is, therefore, imperative that care be taken to ensure the education system is not left behind in streamlining its operations and enhancing systems geared towards producing high-quality performers in a sustained manner.

Understandably, however, the education system has not kept up to date with changing methodologies of management, and although by its very nature the sector receives information on what is trending, helpful or being tested, it often takes quite a while for it to catch up, probably owing to bureaucracy and general conservatism of leaders. Nevertheless, as systems thinkers, it is not for us to decide why such a good approach as systems thinking has not been embraced in the education system. The way to be of use is to conduct a brief analysis of the situation in the education sector and then suggest a sustainable systems-thinking way that the leaders involved can follow to improve the sector.

With the dynamics of society changing and technology becoming extensively used, the education sector has lagged behind other sectors in keeping up to date. This means there are some needs that are within the purview of the education system that are not being met or which are being fulfilled unsatisfactorily. What is the solution? It is to introduce an entirely new systems approach within the education sector, one which will help to address the evolving needs of society even as it addresses the needs of individual students passing through the system.

As has been noted, it is not that the people running the education system are ignorant to the fact that the system is not running as it should be; it is just that they have either not gotten a handle on what the best solution is, or they have been overwhelmed by the complexity of the problems. What the leaders in the sector have done for quite a number of years now is to institute piecemeal solutions, and that has not produced desirable results. What the education system needs right away is total quality management (TQM).

Importance of Total Quality Management (TQM)

In pursuance of total quality management, leadership must, of necessity, seek systems improvement by instituting systemic change. The education sector has largely retained the ancient approaches that were used when systems associated with education were not as complex as they are today and many leaders have opted to let the status quo remain. Often when people who are open to change highlight the need to change the system, they are met with strong opposition, with the many proponents of the status quo arguing that the approach being used in the education system has a record of great success. What they do not mention is that those successes are historical.

What happens when people revel in past successes and ignore the current challenges is that they are often left lagging behind others who are in touch with the reality of a changing environment. In fact, the systems with which the education system interrelates have changed drastically over the last century or so, and that knowledge has been in the public domain for slightly more than half a century. In fact, as Professor Béla Heinrich Bánáthy, who founded the White Stag Leadership Development Program, noted in his 1991 publication, the magnitude of the change in the environment of the education system was only noted in 1950, but the series of reforms that the realization catalyzed appear to have had little impact.

In short, the education sector is still faced with challenges it has no capacity to solve because it is not well equipped in terms of skills and mental attitude. Does this mean stakeholders do not realize there are big challenges facing the education system? No, they are not ignorant of the fact. In fact, there have been persistent calls in recent years for systemic change within the sector and that shows the weaknesses in the sector are big enough and sensitive enough for people to be concerned about. Many people have articulated the need to have changes in the system. One might then ask: why are those changes not made?

For one, any changes would be directed toward the systems within education as well as those that directly relate to it. Think of the way a trading company decides to change its delivery hour. Does it not require that the customer also change what time he or she receives goods in order to work in tandem with the company? When it comes to the problems affecting the education system and the solutions that have the potential to address them, many people involved have a problem defining what exactly constitutes a system, because if they accept that a system is virtually everything, apparently they are not prepared to handle the implications.

In short, while it has been unanimously agreed that there are massive problems in the education system, no one is prepared to have their zone interfered with when it comes to proposals to overhaul the system. It is not even surprising to hear people talking of using systems approach, when in reality their day-to-day operations are deeply rooted in the traditional approach. Many of the stakeholders are yet to appreciate the nature of systems thinking, where a single problem needs to be attacked from different vantage points.

Another challenge is that, owing to the different interpretations people tend to give to the term "system," some involving mechanical models, decision makers often get confused and shy away from supporting systems thinking. For the systems-thinking approach to succeed, it is important for decision makers to clearly understand the reason the education system is still ridden with problems despite numerous attempts at solving them, and also why systems thinking will provide a sustainable solution.

Current Inappropriate Approach to Education

One reason it is so clear the current US education system is failing is that people have had a taste of a successful education system—the same system when it really worked. They know how students spoke in the presence of elders, what the state of security was in educational institutions, what the students' performance was, how healthy students generally were, what the social environment was like in schools, colleges and universities and so on. That makes comparing those aspects to today's system very easy, and it certainly points to a declining trend.

From the very beginning, public education has been geared towards transmitting basic knowledge as well as cultural values, providing custodial care as well as getting students set for the life they are probably going to lead once they complete their formal education. The most fundamental aspect of formal education is always to prepare students for critical and creative thinking, so that they are well placed to solve problems and make sound decisions.

The evaluation that has been done by systems thinkers and other analysts shows that the public education system has done a good job on the first count. That is why you see so many good debaters on all manner of topics, numerous innovations and other interesting things that clearly show that students can think critically and creatively. Does the system today succeed in producing people who are great at solving problems? The answer on this count is, not to impressive levels, while, regarding students making sound decisions, there is even more room for improvement.

Overall, the education system is doing a lot to help create a stable society, but there is certainly more to successful development than stability. So, as the system is being credited with maintaining stability, it is also being flagged for failing to modify its systems to meet societal expectations that have, for a long time, been changing without being addressed. Professor Bánáthy has highlighted five solid reasons why any endeavors to make effective changes to the education system have consistently failed to meet any significant success.

Challenges That Have Hindered Transition in Education
1) Piecemeal implementation of solutions

Sometimes stakeholders purport to implement changes to the traditional system, but instead of instituting those changes with gusto and focus, they do it in piecemeal fashion and often on an incremental basis.

2) Errors of omission with reference to integrating solution ideas

3) Studying and analyzing the state of education discipline by discipline

4) Adopting a reductionist orientation which, essentially, calls for dismantling of the whole so that analysis is conducted on the constituent parts. Yet it is evident that the behavior of a system is not a sum total of the behavior of its constituent parts. Systems thinking recognizes this, and that is the reason system thinkers analyze the behavior of a system within the context of relationships among smaller systems that comprise the bigger one.

5) Remaining within the confines of the system under focus

What this means is that, because it is the education system being discussed, with a view toward identifying problems within it and to seeking solutions to those problems, then the people involved do not find it necessary to look beyond the education sector. This is tantamount to treating education like a self-sustaining system, whose activities have no outside triggers, consumers or stakeholders; clearly a faulty standpoint.

The reason systems thinking works for the most part is that everyone involved is encouraged to think outside the box, not only creatively but also imaginatively and adventurously.

These five challenges that have contributed to the delay in transforming the US education system to meet modern societal needs are good examples of what is defined as paradigm paralysis, whose alternate term is "mumpsimus." What the US leadership needs to appreciate is that nobody ever succeeded by basing current analysis of a situation on ancient models, even if those models have a history of working pretty well. The reason these models were suitable once is that the models then were not old at that point. Now that a lot has changed, modern models are the best to rely on.

All through this analysis, we have treated the education system as if it is one unitary system, whereas it is one system that is pluralistic in nature, like many others, which, as would be expected, has conflicting goals. It is also true that, although the education system is still stable and discharging its role, several compromises have been made by virtue of using old paradigms in a modern context, and as such, performance has been unsatisfactory. This is paradigm paralysis, and it hinders one from getting a proper view of the system deficiencies that need attention.

When asserting that the US education system has failed to conform to the principles of systems thinking, the presumption is that there are other principles it has been run on that are helping to improve performance.

Schools Operating as Open Systems

Schools function as moderately open systems, their main energy coming from the intellectual as well as financial systems. A school cannot be termed a natural system because it does not operate under a social mandate which normally represents consensus among participants. In fact, a school has different legal mandates that are often conflicting. As such, a lot of energy is expended in trying to sustain relationships within the school and fulfill goals.

Schools are also somewhat mechanistic and less organic, and that can be seen from the reality of the rigidity of their structures. These structures often treat every element the same, creating a great deal of uniformity. Good examples are the lengths of class periods, one particular textbook or publication for every student in the class without any variations, the same number of students in each class and the like.

Schools also happen to have a very short range of goals, and the goals for one student are the same for all students. In this respect, schools can be considered unitary. However, when it comes to means of achieving those goals, there is some latitude and one public school does not have to operate exactly like another.

Something else about schools is that their hierarchy is restricted. When a school is faced with numerous constraints, such as those associated with legislative mandates or pressures of an environmental nature such as drug abuse, poverty, racial prejudices and the like, they tend to become all the more closed as systems, and also mechanistic, restricted and unitary.

If anyone wants to improve the quality of education, they need to improve the educational system design so that it can optimize existing relationships among different elements, even as it enhances relationships between the environment and the educational system. Effectively, what this means is that it is imperative that the closed educational system be modified to become more open and pluralistic, as opposed to remaining closed and unitary. It also needs to function more organically as opposed to mechanistically. And as Professor Bánáthy noted, such an educational system behaves as follows:

- It operates like the complex system it essentially is, as opposed to ignoring the relationships that need to be enhanced for the good of the education system.

- It interacts on a constant basis with environments that are always changing and that coordinate with various other systems within the environment.

- It is capable of coping with constant change and uncertainty, as well as ambiguity in the course of maintaining its capacity to evolve in line with the environment, by way of modifying itself as the environment transforms.

- It has the capacity to exist and deal with change in a creative way, while welcoming complex situations that may also be ambiguous, and not just tolerating them.

- It has capacity to act as a set of organizational learning systems, with the capability to differentiate situations, especially those that require modifications and corrections, as well as those that require changing and redesigning.

- It is designed to seek new purposes and create fresh niches within the environment. It is also capable of self-referencing and self-correction, even as it self-directs and self-organizes. It has the capacity to self-renew.

- It has the capacity to recognize that the ongoing knowledge explosion requires increased specialization and diversification as well as integration, together with generalization, on the basis of a two-pronged approach.

- It has the capacity to increase the volume of information it processes and its processing speed, as well as being able to distribute that information to an even wider range of people and groups. As the same time, it is capable of transforming that information into organizational knowledge.

Systems Impact on Education

Inevitably, the fact that the system the country's education system is using is outdated means it is incapable of meeting the demands of modern-day society. Yet the schools and people involved in the system are all part of societal systems, which in turn work hand in hand with other systems for efficiency and balance. After admitting the inadequacy of the systems currently being used in the education system, one can only conclude that, even if all possible amounts of resources were employed to fine-tune the outdated system, its improved performance would not be significant. There is, therefore, a need to find a solution that will fit the education system so that it can meet the demands of the twenty-first century.

Much research and analysis has been done, and the common findings indicate that the demands of today's world require a response that is participative, rather than dictatorial, looking at the scene from a societal point of view. When looking at it from the perspective of education and management, the appropriate organizational systems will be more purpose-driven than deterministic. The recommended styles are found in the systems-thinking approach, and they can help transform the education system to meet the increased needs of modern-day life.

What the Education System Needs to Change

For the purpose of accommodating the right management style and the best approach to problem-solving, the education system needs to change perspective from the orientation of one-to-many and shift to many-to-one. A good example right within the education system is where we cease to view education as a system where one teacher transmits information to many students, and instead view it as a system where many sources of information provide information to a single student, and one of those sources happens to be the teacher. If you think about it critically, you will appreciate there are really a wide range of sources of information accessible to students today, especially considering how technology has transformed the spread of information.

This transition from one-to-many to many-to-one can be described as a shift from emphasizing instruction to emphasizing actual learning. Looking at the whole scenario from a systems-change perspective, the implications of the concept are massive. In the proposed new system, you can no longer view education as seeking information between two book covers, meaning strictly reading textbooks when seated within the four walls of a classroom, within the system's six designated periods. Thus, experts describe the old system that requires replacement as being a "2-4-6 model."

The Look of the Recommended School System

Once the education system has been changed from the old one that served fewer needs, to a new one that meets the increased demands of a much-changed environment, it will have the capacity to:

- Produce better outcomes.

- Set outcome-based standards.

- Have benchmarks for every one of the standards set, against which individual and program progress will be measured on a continuous basis.

- Conduct assessments by comparing actual performance to benchmarks as opposed to basing them on other students or on feedback.

- Accommodate self-evaluation.

- Employ several forms of assessment to be done by several assessors—what experts call triangulation. This will be done with a view to increasing the validity, as well as reliability, of feedback.

- Have the capacity for immediate intervention.

- Have generative learning, as experts have recommended.

- Accommodate reflective practice, as researchers have recommended.

- Engage instructional design that is balanced, as experts in education have recommended.

- Provide a wide variety of learning structures—some self-directed, others one-to-one, in small groups and lecture form; include field study as well as apprenticeships and mentoring.

- Have school programs running throughout the year.

- Use individual performance to set assignments for learning groups instead of relying on age-grade distinctions.

- Have teams working for long periods that can extend to more than one year, so as to attain a common goal; what experts call intact teams.

- Provide a wide range of information resources through enhanced telecommunications both from school as well as home and also use peer relationships as well as cross-age relationships. Other media will also be used, including video and optical structures involving cooperative learning, libraries, recreation centers, etc.

- Enhance access to information.

- Digitize student information, as well as instructional resources, and make the information and other resources wholly accessible via modern phones.

- Provide e-books.

- Provide multilingual resources.

- Deliver information and learning materials via multi-media, including sound, graphics with room to add text.

- Have an integrated curriculum with instruction and assessment.

- Have small, semi-autonomous teams of six to eight people as subsystems of the education system, forming a hierarchy.

One important point to note is that the recommendations outlined here are not all new, but in the case of the recommended new system, efforts would be made to incorporate them all within it. One can appreciate this after considering the implications of having total quality management in the education sector, which would include a comprehensive systems approach, and, apparently, an entirely new system.

The Need to Revitalize Schools

There have been debates in the public arena discussing the apparent drop in the quality of education in the US and the reasons for it, but one common thread from all knowledgeable analysts is the need to make great changes.

In a report published in *Fortune* magazine, the country's school system was labeled as "most endangered" as an institution. The report highlighted the general concern that has been heightening over time, mostly regarding educational quality. It has been noted that illiteracy levels are on the rise as is the dropout rate. Why does this concern the general public?

When the school system does not turn out a knowledgeable young population, people who can think critically and with problem-solving skills, what is thrown back into society is a chunk of the population that is less productive or sometimes not productive at all.

There has also been hue and cry from several corporations that have experienced a shortage of skilled manpower over the years, and they want to see a more innovative approach to education. It is on that basis that Fannie Mae introduced a ten-year mentor program in one school within Washington, DC, in 1989. It is also along that line of thinking that Sears, as well as United Airlines, with 14 other corporations within Chicago, donated $2 million to establish a model school that would practice school- based management.

Citicorp, on its part, together with Exxon and RJR, have been known to donate money in support of the Coalition of Essential Schools, which is a network involving 50 schools that are dedicated to embracing curriculum reform with a focus on small-group learning.

Role Played by Systems Thinking in Education Success

It is clear from these many innovative educational programs that a different school system is being embraced, one with a radical vision that seeks to manage schools like organizations. Just as managers with corporations are held accountable, teachers, too, will be held accountable in the new system. Here they are held responsible for the performance of their students, with school heads having the power to recommend salary increments for individual teaches and also pinpoint teachers for retirement.

The new education system people want to see is one where emphasis is placed on providing students with critical thinking, rather than hard, dry facts. People want the curriculum to deviate from the traditional lecture teaching format and lean towards learner-directed learning. In short, what experts and the general public are calling for is the introduction of systems thinking into schools.

In Jay Forrester's writings on system dynamics with respect to pre-college education, he pointed out the fact that the US education system has continued to decline in its capacity to meet modern needs. In support of this assertion, he highlights modern-day corporate executives, products of this education system, who are finding it difficult to cope with the complexities of national and international competition; government leaders who do not understand either economic or political change; rising unemployment, a rise in drug culture and other inadequacies.

The small section of the education system which has embraced systems thinking through various innovative programs is doing it as a framework within which to address prevalent problems and improve students' capacity to understand a subject, and also as a tool to effect restructuring so as to create a more effective education system.

These two perspectives are not mutually exclusive, as was expressed by Jim Daniel, a member of the Kentucky Educational Foundation (KEF). He says systems thinking builds a new culture right within the school system, and in due course, the approach of systems thinking may find its way into the classroom to be taught to students.

Sample Schools with Systems Thinking

Instead of waiting for the national government to generate policy that supports the introduction of systems thinking into schools, some states have allowed innovative stakeholders to liaise with particular schools to introduce systems thinking. Probably those local authorities are looking at such initiatives as pilot projects that they can emulate as they spread the approach to other schools. Whatever the reason for allowing these initiatives, the fact is that some are already reporting positive changes.

Orange Grove Junior High School

A case in point is Tucson's Orange Grove Junior High School, where, although the school has not entirely discarded the old system, systems thinking has taken center stage in the school's operations. In short, although the school management system can be termed a hybrid for now, the positive impact of systems thinking is already being felt.

The school has introduced systems thinking in most of its science curriculum, and it is continuing to be implemented across other schools in the Tucson. At Orange Grove Junior High, although teachers continue to grade individual students' work, and the school continues to run one-hour classes, the courses being offered do not have specific titles now such as history, geography or biology. How is it possible, then, to describe what is being learned or studied, one might wonder? What is being learned is described in terms of the topic, such as human studies or marine studies. Effectively, the naming of the subject being studied emphasizes systems orientation— the subjects being systems oriented.

As for classes, they are scheduled in block form, meaning science and social studies can follow one another, and the length of each block can be made shorter or longer. Certain blocks can even be combined to enable students to have a better in-depth understanding of a project in either one or both of the subjects involved.

Owing to flexibility of the scheduling, it is possible for teachers in the school to pursue projects of an interdisciplinary nature. A good example is a class which was conducting a mock legal trial in a bid to learn firsthand the workings of the legal system. The dispute in the case was of a scientific nature, and that provided an opportunity for the students to expand their knowledge, not just in the legal field, but also in the field of science.

Not surprisingly, systems thinking has initiated a dramatic change in both teachers' and students' roles at Orange Grove Junior High. One biology teacher, Frank Draper, explained that students are no longer passive recipients of information, but rather they are active learners. Also under the systems-thinking approach, instead of students being taught science via the traditional approach, they are learning means of acquiring knowledge and using it, whether that knowledge is scientific or otherwise.

As Mr. Draper noted, the role of teachers has changed from being information dispensers to producers of enabling environments, with these environments empowering the students to learn as extensively and deeply as possible.

Technology Experts Help Introduce Systems Thinking

In 1985, when only a few people were equipped with computer skills, the Systems Thinking and Curriculum Network Project, otherwise referred to as STACI, started its work at Brattleboro Union High School, in Vermont. At the school, four teachers were involved because they wanted to enhance their teaching using systems thinking, but they were not computer literate and did not have any knowledge of systems theory. Instead of letting their lack of computer skills discourage them, they applied for a grant from the US Department of Education and received it. They were then able to engage the services of High Performance Systems for training and technical support. It was while trying to acquire the relevant skills that their endeavors drew the attention of Ellen Mandinach, a researcher with the Educational Testing Service or ETS.

Mandinach's interest lay in understanding how technology impacted learning. He therefore organized a liaison between ETS and STACI, so they could examine how things went as systems thinking was introduced in Brattleboro. Organizing this kind of study was probably made easier by the fact that ETS was affiliated with the Educational Technical Center at Harvard.

Apple then came into the picture, with one of its representatives prompting ETS to initiate a similar pilot project on the West Coast. The project on the West Coast benefitted from lessons learned in

Vermont, particularly that teachers were in need of immense resources as well as training and support. So, as Mandinach later explained, instead of picking just one site for this initiative within California, they opted to engage a consortium of schools. This would ensure there was a continuity of systems learning where students were concerned. For starters, they identified two middle schools in San Francisco and four high schools associated with those middle schools.

Students Tackle Real-Life Problems

What remains now is to demonstrate that systems thinking does have the capacity to change students from exam-focused individuals to people who are well equipped and psychologically prepared to tackle real-life problems.

When systems-thinking initiatives were begun in a couple of schools, there might have been skepticism among some people on their capacity to teach young students any better than the old teaching approach did. It must, therefore, have been gratifying for the people involved to see tangible results from their initiatives.

The Orange Grove Middle School systems-thinking initiative began a project dubbed New State Park. As Mr. Draper explained, students were charged with conducting research on park philosophy and park management, as well as land management, geography and ecological community theory, recreation theory, politics and even social systems.

After being equipped with that wealth of knowledge, they were supposed to design an entirely new park on a budget of $100 million. That park needed to factor in certain limitations and risks while utilizing land as required by the park's charter. Some of the sensitive issues the students were required to take into account included the possibility of being sued if they happened to desecrate an Indian burial site that was close by. They were also supposed to ensure the park they designed was attractive to potential users even as the development did not cause any appreciable environmental degradation.

The students involved in the project did their project design on computer, and then they made use of a spreadsheet to document spending for the sake of fiscal accountability. It was also important to keep records for the sake of design accountability, as per the STELLA model with respect to park development and environmental degradation.

The students were expected to present their final design to other students, and through all their activities, presentation included, they developed mental models, with the help of the teachers, of how to go about deciding how to make good use of the land provided. At the end of the day, they were able to make connections involving facts which, at a casual glance, appeared interrelated. In short, by employing systems thinking in teaching, students were able to acquire learning and skills that many adults hardly have a chance to acquire.

At the same time the students demonstrated they could think critically and undertake a project of value to the community, they proved they could take care not to infringe on other people's rights. In short, it was clear that with the skills acquired through systems thinking, the students could cope with modern societal demands.

Peter Senge, who was also interested to see the impact of systems thinking in schools, visited Orange Grove Middle School, and the students involved with the project were eager to show him the plan they had drawn up for the proposed national park. A debate developed, with Senge suggesting to the students that the park could do with more hiking trails than they had provided. Senge, making his case, argued that the hiking trails would attract more visitors and therefore boost revenues. The students countered Senge's suggestion by explaining that adding hiking trails would end up lowering the park's environmental index, a bad thing.

Senge also wanted to know how they felt about having an Indian burial ground within the park area, knowing the burial ground was a respected place for the Indian community, yet liberal activities would take place in the park. In response, the students suggested construction of an information center to showcase Indian culture, very likely winning over the Indian community and showing them goodwill.

It only took Senge 20 minutes to be convinced the students had gained immense knowledge on a national park's economic-social-ecological system; in fact, more than he knew. He was also impressed to know that they were also open to new ideas, because even as they explained the basis for their decisions, they promised to reconsider their stance on the economic projections they had made. When Senge later learned that the two students who had led him through the park plan had been disciplinary cases under the old school system, he felt vindicated for having advocated systems thinking as a way of solving various social problems

Why Formal Training Might Help

People have heard a lot about the goodness of systems thinking and how it can turn around the fortunes of an organization, be it for profits or just efficiency in service delivery. However, because it is a relatively new approach to management, it has not been incorporated in most curriculums at any level. As such, company executives, project managers and everyone else interested in systems thinking has largely learned either on the job or been self-taught through books. It is probably time to admit that formal training is necessary so that those institutions interested in using systems thinking stop facing unnecessary resistance from leaders of other systems who do not understand it.

Already, the MIT Sloan School of Management has begun exploring how best to introduce systems thinking to the workplace, and it has established a program for this purpose. The program is referred to as the "Systems Thinking and Organizational Learning Research Program," and one part of it deals with research. That research part has two main objectives, one of the objectives being to craft a course geared towards teaching a range of skills in systems thinking. It is also meant to assess the effectiveness of systems thinking once adopted into the organization, mainly enhancing integration between systems thinking and corporate decision-making.

The project being undertaken by MIT is "The Systems Thinking Competency Course" (STCC), and it is a collaboration between various corporations and academia, where researchers from MIT will work closely with corporate sponsors with a view toward defining the ideal scope of the project and what its content should be.

The manager in charge of the project, Janet Gould, has highlighted the three main areas the project seeks to address. She says the areas seek to address the following questions:

1) What is the meaning, in the real sense, of being competent in systems thinking?

2) What are the actual skills that people need to acquire before they can be considered to be competent in systems thinking?

3) What are the complementary skills that people need to acquire in order to become great facilitators of the systems-thinking approach in an organization?

For starters, a lot of work has been put into defining the items that are important to include in the systems-thinking competencies list.

In the diagram below, you can see the proposed framework designed to address the issue. It is important for the research to decide the particular skills to be included in each matrix cell. At the beginning the team cannot say with certainty that it is not going to make alterations to these entries, but it is a good start.

Although discussions are still ongoing regarding course content, there are a few formats for content delivery that have been suggested. Among the suggestions is conducting a five-day intensive course, and then, with enough participants, delving right into the content and teaching the principles of systems thinking.

One of the project's sponsors, Hanover Insurance Company's Tom Grime, says an experience like that of intense learning is likely to help participants actually retain what they will have learned in the short, intensive course. He explains that people's capacity to think in a linear manner is so vast that there is a need to take people through an intense learning experience in order to change their mentalities.

	A Novice	An advanced Beginner	One who is competent
Tools to use: Raw materials Templates STELLA	Understanding Gaining awareness N/A	Applying Understanding and recognizing Exposure	Adapting Internalizing Understanding STELLA model
Principles	Gaining exposure	Understanding and recognizing	Applying
Applications	Gaining exposure	Understanding	Application of 1&2 to own problems

The project still has another format, entailing three full days of instruction, and after that some refresher courses will follow, held once in every couple of months. In the meantime, the participants can be asked to maintain logs showing how the systems-thinking approach seems to be affecting their work. In this format, the iterative process can go on for at least one year.

304

What the project emphasizes, regardless of the format the training takes, is learning as a team. A recommendation has been made encouraging several people from one division to undertake the training as a group, and then they can continue to apply the skills they learn together when they resume work at their workstations. When people from the same department or section undergo the training together, they will be in a good position to compare notes on a daily basis as their proximity to one another makes that convenient.

Although the formal training is planned for short periods over the course of a few days, Gould notes that the participants will effectively be training for longer, considering that, as they apply those skills and talk about their effect, they will still be learning. He says this is much better than having individuals from different divisions of the organization attending the training together, because the latter would not have time to exchange related ideas and learn more together.

MIT has taken the issue of systems-thinking training seriously, and the institution has already decided how to model the course. There will be a one-week long summer session geared towards introducing participants to the systems-thinking approach, and another five-day course that was crafted by consultant David Kreutzer specifically for his clientele.

Both of those courses intend for participants to learn feedback loops, at least those that are simple to understand, and also how to create simple computer models to represent complex systems. Naturally, this course is not, by any means, exhaustive. In fact, there are a number of cells in the research matrix that have not been filled. Nevertheless, in a country where people are still not sure how to begin implementing systems thinking, the course is a good start.

Objectives of Systems-Thinking Beginners Course

As a group, participants can learn a lot because there is room to interact, exchange notes and ask questions, but for the official part, there are specific objectives that Tom Grime wants the course to fulfill, as listed here:

(1) Make the participants aware that linear thinking has many limitations, and then tell them what the main limitations are.

(2) Make the participants aware that linear thinking poses potential dangers, and then point out some of those potential dangers.

(3) By using systems thinking, help the participants pinpoint various assumptions that people make to justify their decisions.

(4) Provide participants with one common language that they can use when expressing issues relating to systemic issues.

(5) Provide participants with a forum within which to critique people's general view of reality, and expand on it, while avoiding any temptation to become emotional or personal.

MIT's plan is to have the course tried out by four or five companies. Although MIT has been involving companies in these pilot courses, it is solely for the purpose of research. Thereafter, even as those participating individuals and companies benefit, the researchers will be in a position to write a report that is backed with credible findings. Already, Project Manager Gould has mentioned that the researchers need to learn whether the things they have assumed to be important in teaching systems thinking are really working as anticipated.

However, the aspect of team learning in the course cannot be underrated. It needs to be remembered that one of the reasons the linear approach does not do well in an organization is the tendency for people to apportion blame when things do not work well. With team participation, it is easy to enhance relationships among team members, so that, instead of trying to hunt for someone to blame, people share information as they try to identify the part of the system that is the source of the problem.

However the MIT training goes, or others that might be held in the future, one thing course planners agree on is the fact that systems thinking is an extremely valuable tool, and organizations that adopt it are bound to experience real improvement in company the work environment and efficiency of daily processes, as well as sustained growth.

Systems-Thinking Approach in Summary

There is a lot to say about systems thinking and that is mainly because it deals with the betterment of virtually everything. After all, everything is a system. From an atom in the body to the vast waters of the ocean, all are systems that have a relationship with one another. After reading this book and understanding the benefits of systems thinking and how it is used to enhance relationships and the sustained efficiency of systems, you may wish to think of systems thinking as a means to help you succeed in a specific area.

Company executives might be interested in enhancing relationships at the workplace, so that members of each department can understand that none of them can be successful or happy unless they develop great relationships with other departments. Individuals might wish to look at how internalizing the principles of systems thinking can help improve their personal relationships. In short, although everyone and everything is a small system that is a member of the ultimate big system, the universe, you need to know what you are seeking and what, specifically, you are seeking to achieve in order to identify the systems to involve in your endeavors.

The simplest advice given by experts when you want to engage in systems thinking is to first examine your entire immediate system, for example, your company, your family, your local supermarket, the city bus system and the like. In the jargon of systems thinking, this kind of overview is termed "getting yourself up high in the helicopter."

As you look from above, mentally, of course, you are trying to see what is actually inside your system, and what is also on the outside. Those are the subsystems that are likely to affect your processes, and it is necessary that you ensure your system works in harmony with them. The relationships you have with each one of those systems need to be seamless and harmonious.

Since systems thinking is nothing like the linear approach, where you have a cause-and- effect situation every time, it means you are called to examine different relationships at every phase of the process because you are open to discovering the problem or bottleneck at any location or itcration. There are some questions that can help you ensure you are approaching systems thinking the right way:

Systems-Thinking Questions for Every Phase

Phase 1: Where, exactly, do we want to reach? (seeking the ideal future)

Phase 2: How shall we tell that we have arrived? (seeking measure of success)

Phase 3: Where, precisely, are we? (seeking to know current state)

Phase 4: How shall we reach where we are aiming? (seeking means of implementation)

Phase 5: What is likely to change within the environment at a later time? (seeking ongoing progress associated with external factors)

Conclusion

Believe it or not, systems thinking in fact touches every part of our life in one way or another. If we are not part of it, some others will surely be part of it in an effort to improve themselves.

I hope you enjoyed and benefited from this book. The next step is to share what you have learned.

If you did enjoy this book, please do leave a positive review for the book on Amazon.

Thank you for reading this book.

Made in the USA
San Bernardino,
CA